Lecture N
Mathematics

Edited by A. Dold and B. Eckmann

418

Localization in Group Theory and Homotopy Theory

and Related Topics

Edited by Peter Hilton
Battelle Seattle Research Center 1974

Springer-Verlag
Berlin · Heidelberg · New York 1974

Prof. Dr. Peter J. Hilton
Battelle Seattle Research Center
4000 N.E. 41st Street
Seattle, WA 98105/USA

Library of Congress Cataloging in Publication Data

Main entry under title:

Localisation in group theory and homotopy theory, and related topics.

(Lecture notes in mathematics ; 418)
Papers presented at a symposium held at the Battelle Seattle Research Center, Mar. 11-15, 1974.
Bibliography: p.
Includes index.
1. Groups, Theory of. 2. Homology theory. 3. Homotopy theory. I. Hilton, Peter John, ed. II. Battelle Memorial Institute, Columbus, Ohio. Seattle Research Center. III. Series: Lecture notes in mathematics (Berlin) ; 418.
QA3.I28 no. 418 [QA171] 510'.8s [514'.23] 74-22375

AMS Subject Classifications (1970): 20D15, 20J05, 55D15, 55D45

ISBN 3-540-06963-1 Springer-Verlag Berlin · Heidelberg · New York
ISBN 0-387-06963-1 Springer-Verlag New York · Heidelberg · Berlin

Offsetdruck: Julius Beltz, Hemsbach/Bergstr.

FOREWORD

In the four years since Sullivan first pointed out the importance of the method of localization in homotopy theory, there has been considerable work done on further developments and refinements of the method and on the study of new areas of application. The Battelle Seattle Research Center therefore decided to act as host to a symposium on localization theory, which took place at the Center during the week of March 11 - 15, 1974. The opportunity was a particularly good one, since at that time, Guido Mislin and Joseph Roitberg were at the Center as Visiting Fellows.

Since the technique of localization has involved the study of purely group-theoretical aspects of the theory, and since, in its general aspects, it is related to abstract work in homology theory and category theory, it was decided to make the terms of reference of the symposium fairly broad so that the many facets of localization theory could receive attention. Thus, although all the papers in this record of the conference are definitely related to localization, they are not necessarily concerned with the theory itself, nor with its special application to homotopy theory.

It is a pleasure to acknowledge the kindness of many people at the Battelle Seattle Research Center who helped to make the symposium a pleasant and productive one. Among many deserving mention are Ms. Evelyn Zumwalt, who was responsible for arranging the hospitality for the participants; Mss. Julie Swor and Shirley Lake, who were responsible for the organizational aspects of the symposium; and Ms. Sandra Smith, who prepared many of the manuscripts. To all of these and many others, I would like to express the deep appreciation of all those who attended the symposium and derived great benefit from it.

Battelle Seattle Research Center
June, 1974

Peter Hilton

CONTENTS

CONVERGENT FUNCTORS AND SPECTRA

by D. W. Anderson

There are two homotopy theoretic constructions which have been used to describe homology theories: spectra (topological as in Whitehead [1962] or simplicial as in Kan [1963]) and special Γ-spaces as described by Segal [1970] and as used by Anderson [1971]). While both spectra and special Γ-spaces have certain advantages not shared by the other, spectra continue to be used because they have a certain amount of flexibility which special Γ-spaces do not have--the category of spectra has finite colimits, and there is a notion of a pairing of spectra (though the old problem of defining an associative, commutative smash product for spectra still eludes solution).

In this talk, I shall outline a third description of homology theories on topological spaces which is similar to my use of special Γ-spaces to describe homology theories on simplicial sets. Rather than special Γ-spaces we shall consider certain functors from pointed spaces to pointed spaces which we call convergent functors. The category of convergent functors will have finite limits and colimits, just as the category of spectra does, but furthermore it will admit a naturally associative and commutative smash product. Thus, the category of convergent functors promises to be a more useful tool for studying delicate homotopy theoretic properties of homology theories than spectra have been.

To begin, before I define the term "convergent functor", I shall give an example of a convergence functor which has been known for some time. If X is a pointed topological space, let $SP(X)$ be the infinite symmetric product of X. Then, at least if X is a CW-complex, the ordinary integral homology groups of X can be obtained from $SP(X)$ by the relation $\tilde{H}_n(X;Z) = \pi_n SP(X)$ (see Dold-Thoms [1956]).

Suppose now that Φ is a functor which assigns to every pointed space another pointed space. If Φ satisfies certain conditions given below, we shall call Φ a convergent functor, and if Φ satisfies one further condition which we call additivity, we call Φ a convergent chain functor. When Φ is a convergent chain functor, $H_*(X) = \pi_*\Phi(X)$ will satisfy the Eilenberg-Steenrod axioms (except the dimension axiom for a homology theory), at least if X is restricted to the category of CW-complexes.

Suppose now that we have any functor $\Phi: \mathcal{T} \to \mathcal{T}$, where $\mathcal{T}$ is the category of pointed spaces. If $\pi_*\Phi$ is to be a homology theory, it must satisfy the homotopy axiom. One way to insure this is to assume that we have a natural transformation

*The author was partially supported by NSF Grant GP-34489.

$\Phi(X) \wedge K \to \Phi(X \wedge K)$ which is defined for X a pointed space and K a polyhedron (by which we shall mean the geometric realization of a finite simplicial set), which is the identity for K the zero-sphere, and which is associative with respect to K. A functor Φ provided with such a natural transformation will be called a simplicial functor. Notice that if $F: X \wedge I^{+} \to Y$ is a homotopy, where I^{+} is the one point compactification of the unit interval, the composition

$$\Phi(X) \wedge I^{+} \longrightarrow \Phi(X \wedge I^{+}) \longrightarrow \Phi(Y)$$

is a homotopy. Thus we see that if Φ is simplicial, Φ carries homotopic maps to homotopic maps.

If $\mathfrak{M}$ is a spectrum with $n^{\underline{th}}$ term M_n, define a functor $\mathfrak{M}$ by letting $\mathfrak{M}(X)$ be the mapping telescope of the sequence $M_0 \wedge X \to \Omega(M_1 \wedge X) \to \Omega^2(M_2 \wedge X) \to \ldots$. Then $\mathfrak{M}$ is clearly a simplicial functor, and $\pi_*\mathfrak{M}(X)$ is Whitehead's $\tilde{H}_*(X;\mathfrak{M})$. Thus any homology theory defined in terms of a spectrum can also be defined in terms of a simplicial functor.

If Φ is a simplicial functor, let $\Phi_n(X)$ be $\Omega^n(\Phi(X \wedge S^n))$, where S^n is the n-sphere. Then $(\Phi_n)_1 = \Phi_{n+1}$ for all n, and the map $\Phi(X) \wedge S^1 \to \Phi(X \wedge S^1)$ induces natural transformations $\Phi = \Phi_0 \to \Phi_1 \to \Phi_2 \to \ldots$. If we let $\Phi_\infty(X)$ be the mapping telescope of the $\Phi_n(X)$, we can define $\tilde{H}_i(X;\Phi) = \pi_i\Phi_\infty(X)$. As Φ_∞ is a simplicial functor, $H_*(-;\Phi)$ satisfies the homotopy axiom for a homology theory whenever Φ is a simplicial functor.

We now shall give conditions which will insure that $H_*(-;\Phi)$ will be a homology theory. In order to simplify our exposition, we shall assume that all spaces have the homotopy type of a CW-complex. A more general but more complicated exposition will appear elsewhere.

If Φ is a simplicial function, and if $X_0 \to X_1 \to \ldots$ is a sequence of spaces and maps, there is an obvious map of the mapping telescope $\mathrm{Tel}\{\Phi(X_i)\}$ into $\Phi(\mathrm{Tel}\{X_i\})$ coming from the maps $\Phi(X_i) \wedge I^{+} \to \Phi(X_i \wedge I^{+})$, where I is the interval. We call Φ continuous if $\mathrm{Tel}\{\Phi(X_i)\} \to \Phi(\mathrm{Tel}\{X_i\})$ is always a homotopy equivalence. Clearly given any spectrum $\mathfrak{M}$, the functor $\mathfrak{M}$ is continuous.

For any simplicial space X_* one may form a geometric realization $|X_*|$ by making suitable identifications along faces of the union of the spaces $X_n \wedge (\Delta_n^{+})$, where Δ_n is the standard n-simplex (we do not make identification along degeneracy maps to avoid certain complications). The maps $\Phi(X_n) \wedge (\Delta_n)^{+} \to \Phi(X_n \wedge (\Delta_n^{+}))$ give us a map $|\Phi(X_*)| \to \Phi(|X_*|)$. We shall call Φ a geometric functor if this map is always a homotopy equivalence. It is easy to see, using standard results about geometric realizations of fibrations, that if $\mathfrak{M}$ is a connected spectrum (each M_n is (n-1)-connected), $\mathfrak{M}$ is a geometric functor.

We call Φ a convergent functor if it is a continuous geometric simplicial functor. Convergent functors can be constructed from continuous functors from basepointed sets to spaces as follows. If Λ is a functor from basepointed sets to basepointed spaces, define Λ' as follows. If X is a space, let $\mathrm{Sing}(X)$ be the simplicial set which is the singular complex of X. Let $\Lambda'(X) = |\Lambda \mathrm{Sing}(X)|$. An example of such a construction is given by letting Λ assign to a set E the free abelian group on E, thought of as a discrete topological space. One can show that in this case $\tilde{H}_*(X;\Lambda') = \pi_*\Lambda'(X)$ is the singular homology of X.

Our main theorem is the following.

Theorem 1. If Φ is a convergent functor, $H_*(-;\Phi)$ satisfies the Eilenberg-Steenrod axioms for a homology theory (except for the dimension axiom).

Before we prove Theorem 1, we introduce a further definition which is reminiscent of Segal's "specialness" condition for Γ-spaces. We call a functor Φ additive if for all X, Y, $\Phi(X \vee Y) \to \Phi(X) \times \Phi(Y)$ is a homotopy equivalence. If Φ is additive, notice that the folding map $X \vee X \to X$ induces on $\Phi(X)$ the structure of a homotopy abelian H-space. We say that Φ is complete if each $\Phi(X)$ has a homotopy inverse--that is, if each $\pi_0\Phi(X)$ is a group.

For the present, we observe that if Φ is additive, $\Phi(X) \to \Phi(X \vee Y) \to \Phi(Y)$ is a homotopy theoretic fibration.

Proposition 2. If Φ is geometric, Φ_∞ is additive.

Proof. Let $\Phi(X,Y)$ be the mapping cone of $\Phi(X \vee Y) \to \Phi(X) \times \Phi(Y)$. Since, up to homotopy equivalence, geometric realization preserves unions and products, $\Phi(-,-)$ preserves geometric realizations in both variables up to homotopy equivalence. Since the n-sphere, and hence the n-fold suspension of any space in the geometric realization of a simplicial space which consists of just the basepoint in degrees less than n, this implies that $\Phi(S^iX,S^jY)$ is $i+j-1$ connected. Thus $\Phi(S_nX \vee S^nY) \to \Phi(S^nX) \times \Phi(S^nY)$ is a $(2n-1)$-homotopy equivalence. Thus $\Phi_n(X \vee Y) \to \Phi_n(X) \times \Phi_n(Y)$ is an $(n-1)$-equivalence.

Proposition 3. If Φ is a complete additive convergent functor, then for any map $f\colon A \to B$ of spaces, the sequence $\Phi(A) \to \Phi(B) \to \Phi(\tau(f))$ is a homotopy theoretic fibration.

Proof. Let $M_*(f)$ be the simplicial space which in degree n is $B \vee A \vee \dots \vee A$ $(n+1)$ copies of A, with face maps determined by folding maps (analogous to the bar construction). Then there is an augmentation $M_0(f) \to B$ together with an evident contracting homotopy $B \to M_0(f) \to M_1(f) \to \dots$, so that there is a homotopy equivalence $|M_*(f)| \to B$. Further, the map $A \to M_n(f)$ which includes A as the last cofactor induces a cofibration $|A_*| \to |M_*(f)|$, where A_* is the constant simplicial space which is A in each degree. Since $|A_*| \to A$ is a homotopy equivalence, $|M_*(f)/A_*| \to T(f)$ is a homotopy equivalence. Notice that $A_n \to M_n(f) \to T_n(f) = M_n(f)/A_n$ is a split cofibration, so that $M_n(f) = A_n \vee T_n(f)$.

Thus for each n, $\Phi(A_n) \to \Phi(M_n(f)) \to T_n(f)$ is a homotopy theoretic fibration. Since these are all complete H-spaces and all maps in question are H-maps, the geometric realization $|\Phi(A_*)| \to |\Phi(M_*(f))| \to |\Phi(T_*(f))|$ is also a homotopy theoretic fibration. The fact that Φ is geometric implies that $\Phi(A) \to \Phi(B) \to \Phi(T(f))$ is a homotopy theoretic fibration.

Notice that proposition 3 shows that $H_*(-;\ )$ satisfies the exactness axiom, and thus is a homology theory. The proof of proposition 3 can be modified slightly, to prove the following result. The proof is left as an exercise.

Proposition 4. If Φ is a complete linear convergent functor, the maps $\Phi(X) \to \Phi_1(X) \to \dots$ are all homotopy equivalences. Thus, in this case, $\pi_*\Phi(X) = H_*(X;\Phi)$.

If Φ, Ψ are two convergent functors, define $\Phi \otimes \Psi$ to be the functor of two variables defined by $(\Phi \otimes \Psi)(X,Y) = \Phi(X) \wedge \Psi(Y)$. If Λ is a third convergent functor, by a pairing $(\Phi,\Psi) \to \Lambda$ we shall mean a natural transformation $\Phi \otimes \Psi \to \Lambda\mu$, where $\mu(X,Y) = X \wedge Y$. We also require this natural transformation to be compatible with all simplicial structures. Equivalently, we could define a pairing to be a natural transformation $\Phi \wedge \Psi \to \Lambda$, where $\Phi \wedge \Psi = L^{\mu}(\Phi \otimes \Psi)$ is the left Kan extension of $\Phi \otimes \Psi$ along μ. The properties of the left Kan extension immediately imply that the smash product of convergent functors is naturally associative and naturally commutative. Standard arguments can be used to show that the smash product of convergent functors is again convergent.

As things have been done, the identity functor is not a strict unit for the smash product. It would be if either we restricted our category to finite complexes or if we assumed that the natural transformations $\Phi(X) \wedge K \to \Phi(X \wedge K)$ extended to all topological spaces K. Either assumption complicates certain arguments, but such variations of the general theory can be carried out.

If we have a pairing $\Phi \wedge \Psi \to \Lambda$, we can define $\tilde{H}_i(X;\Phi) \wedge \tilde{H}_j(Y;\Psi) \to \tilde{H}_{i+j}(X \wedge Y;\Lambda)$ as follows. We have for all m, n maps $\Phi(X \wedge S^m) \wedge \Psi(Y \wedge S^n) \to \Lambda(X \wedge Y \wedge S^{m+n})$ and thus a pairing $\pi_{i+m}\Phi(X \wedge S^m) \otimes \pi_{j+n}(Y \wedge S^n) \to \pi_{i+j+m+n}(X \wedge Y \wedge S^{m+n})$, which in the limit give the required tensor product of homology classes.

Cohomology groups $H^n(X;\Phi)$ can be defined as the direct limit of the homotopy classes of maps of $X \wedge S^i$ into $\Phi(S^{i+n})$ as i increases. Clearly a pairing of spectra also induces a pairing of cohomology theories.

There is a second type of pairing for convergent functors. Clearly, the composition of two convergent functors is again convergent. By a composition pairing of Φ and Ψ into Λ, we simply mean a simplicial natural transformation $\Phi\Psi \to \Lambda$. A composition pairing defines a pairing of the previous type, at least if the simplicial structure extends to a sufficiently large category of spaces by the composition:
$\Phi(X) \wedge \Psi(Y) \to \Phi(X \wedge \Psi(Y)) \to \Phi\Psi(X \wedge Y) \to \Lambda(X \wedge Y)$.

It is easy to define slant products between homology and cohomology if one has a composition pairing of functors.

Biography:

Anderson; D. W.: Chain Functors and Homology Theories, Lect. Notes in Math (Springer) v. 249 (1971), 1-12.

Dold; A. and Thom, R.: Une Généralisation de la notion d'espace fibré. Application aux produits symétriques infinis, C. R. Acad. Sci. Paris, v. 242 (1956), 1680-1682.

Kan, D. M.: Semi-simplicial Spectra, Ill. J. Math., V. 7 (1963), 463-478.

Segal, G.: Homotopy Everything H-spaces (preprint) (1970).

Whitehead, G. W.: Generalized Homology Theories, Trans. AMS (1962), v. 102, n. 2, 227-283.

THE GENERALIZED ZABRODSKY THEOREM

Martin Arkowitz

Dartmouth College

1. THE MAIN RESULT

This note is based on Zabrodsky's paper [11]. We work in the category of pointed topological spaces of the homotopy type of connected CW-complexes. Mappings and homotopies are to preserve base points and the relation of homotopy is denoted by "$\simeq$". The weak pullback of maps $f: X \to A$ and $g: Y \to A$ is the space $P = \{(x,\lambda,y) | x \in X, \lambda \in A^I, y \in Y, f(x) = \lambda(0), g(y) = \lambda(1)\}$. The maps $r: P \to X$ and $s: P \to Y$ defined by $r(x,\lambda,y) = x$, $s(x,\lambda,y) = y$ are called the projections. The following definitions will be useful.

Definitions. If (Y,μ_Y) is an H-space, A a space and $g: Y \to A$, then an operation of Y on A is a map $\mu_0: Y \times A \to A$ such that $\mu_0|Y \simeq g$, $\mu_0|A \simeq 1$ and $g\mu_Y \simeq \mu_0(1 \times g): Y \times Y \to A$ (fig. 1). If in addition (A,μ_A) is an H-space and μ_0 is an operation, then $\mu_A(g \times 1)$ and μ_0 are homotopic on $Y \vee A$. Hence there exists a map $w: Y \wedge A \to A$, unique up to homotopy, such that $\mu_0 \simeq wq + \mu_A(g \times 1)$, where $q: Y \times A \to Y \wedge A$ is the projection onto the smashed product and "+" denotes addition of maps into A using the multiplication μ_A. We call w the difference element of the operation μ_0.

We now state our main theorem. It is a generalization of a result which is embedded in [11].

Theorem GZ. Given H-spaces (X,μ_X), (Y,μ_Y) and (A,μ_A), a map $g: Y \to A$ and an H-map $f: (X,\mu_X) \to (A,\mu_A)$. Let $\mu_0: Y \times A \to A$ be an operation of Y on A with difference element $w: Y \wedge A \to A$. Further, assume there exists a map $\tilde{w}: Y \wedge X \to X$ such that $f\tilde{w} \simeq w(1 \wedge f)$ (fig. 2). If P denotes the weak pullback of f and g with projections $r: P \to X$ and $s: P \to Y$ (fig. 3), then P is an H-space and s is an H-map. If, in addition, $\tilde{w}$ is nullhomotopic, then r is an H-map.

Research supported by NSF Grant GP29076A2

$$\begin{array}{ccc} Y \times Y & \xrightarrow{\mu_Y} & Y \\ \downarrow{\scriptstyle 1\times g} & & \downarrow{\scriptstyle g} \\ Y \times A & \xrightarrow{\mu_O} & A \end{array} \qquad \begin{array}{ccc} Y \wedge X & \xrightarrow{\tilde{w}} & X \\ \downarrow{\scriptstyle 1\wedge f} & & \downarrow{\scriptstyle f} \\ Y \wedge A & \xrightarrow{w} & A \end{array} \qquad \begin{array}{ccc} P & \xrightarrow{s} & Y \\ \downarrow{\scriptstyle r} & & \downarrow{\scriptstyle g} \\ X & \xrightarrow{f} & A \end{array}$$

(figure 1) (figure 2) (figure 3)

We will not give the proof since it just an adaptation of Zabrodsky's argument to this more general situation. However, we make a few remarks.

Remarks. (a) If $g: (Y,\mu_Y) \to (A,\mu_A)$ is an H-map then we take $\mu_O = \mu_A(g\times 1)$, and it follows that μ_O is an operation. Hence $w \simeq 0$, the constant map, and so we can set $\tilde{w} = 0$. Therefore we retrieve the familiar result that the weak pullback of H-spaces and H-maps is an H-space and the projections are H-maps. Theorem GZ can thus be regarded as an asymmetric version of this result.

(b) The proof of the theorem shows that there is an operation of P on X. Thus if there exists an H-map $h: (Z,\mu_Z) \to (X,\mu_X)$ and if the condition on the difference element is satisfied, then it is possible to iterate the procedure and conclude that the weak pullback of r and h is an H-space with horizontal projection an H-map.

(c) There is a special case which frequently arises where the condition on $\tilde{w}$ is always satisfied. Let ΣB denote the reduced suspension of the space B and let $\theta_k: \Sigma B \to \Sigma B$ denote the map which is k times the identity map, where addition of maps is obtained from the suspension structure of the domain. Now assume the following variation of the hypotheses of Theorem GZ holds: There are maps $f: X \to A$ and $g: Y \to A$, where (Y,μ_Y) is an H-space, and there exists an operation μ_O of Y on A. Also suppose there exists a multiplication μ_X on X and a multiplication μ_A on A such that $f: (X,\mu_X) \to (A,\mu_A)$ is an H-map. Then μ_A and μ_O give rise to a difference element $w: Y \wedge A \to A$. Furthermore, assume (1) $X = A = \Sigma B$, for some space B (2) $f \simeq \theta_k: \Sigma B \to \Sigma B$, for some integer k. With these hypotheses we can show the existence of $\tilde{w}$ such that $f\tilde{w} \simeq w(1\wedge f)$. We set $\tilde{w} = w$ and identify $Y \wedge X = Y \wedge A = Y \wedge \Sigma B$ with $\Sigma(Y\wedge B)$. Under this identification $1 \wedge f: Y \wedge X \to Y \wedge A$ corresponds to $\theta_k: \Sigma(Y\wedge B) \to \Sigma(Y\wedge B)$. Thus it suffices to show the composition $\Sigma(Y\wedge B) \xrightarrow{w} \Sigma B \xrightarrow{\theta_k} \Sigma B$ is homotopic to the composition $\Sigma(Y\wedge B) \xrightarrow{\theta_k} \Sigma(Y\wedge B) \xrightarrow{w} \Sigma B$. But $\theta_k: \Sigma B \to \Sigma B = A$ is homotopic to k times the identity map, where addition of maps is now obtained from the multiplication μ_A. This is so because the two group operations in the set of homotopy classes $[\Sigma B, A]$ - the one obtained from the suspension structure of ΣB and the other from the H-structure of A - coincide. Thus $\theta_k w$ and $w\theta_k$ both represent the element $k[w]$ in $[\Sigma B, A]$ and are therefore homotopic. This shows $f\widetilde{w} \simeq w(1 \wedge f)$ and therefore establishes the hypotheses of Theorem GZ in this special case.

In the following sections we will show how Theorem GZ can be applied to construct H-spaces and H-maps. The two main methods for constructing new or exotic H-spaces are (i) Zabrodsky's theorem [11] (ii) the theorem of Harrison-Stasheff [3,8] and Mimura-Nishida-Toda [5] which we hereafter refer to as the HSMNT theorem. The proofs of (i) and (ii) are quite different. In §2 we derive a proposition from Theorem GZ which implies the HSMNT theorem. This shows that both methods for obtaining new H-spaces are consequences of Theorem GZ. In §3 we sketch a proof of Zabrodsky's theorem by localization methods and also obtain some information on H-maps between new H-spaces.

2. THE HARRISON-STASHEFF-MIMURA-NISHIDA-TODA THEOREM

We denote by $v_2(k)$ the exponent of 2 in the prime decomposition of the integer k and by $\underline{k}: S^n \to S^n$ the map of degree k of the n-sphere. Let H be a closed subgroup of the connected Lie group G and suppose the homogeneous space G/H is S^n. Then $H \to G \xrightarrow{\pi} S^n$ is a principal H-bundle and we let E_k denote the principal H-bundle over S^n induced from π via $\underline{k}$:

$$\begin{array}{ccc} E_k & \xrightarrow{s} & G \\ \downarrow r & & \downarrow \pi \\ S^n & \xrightarrow{\underline{k}} & S^n \end{array} .$$

<u>Proposition</u>. With the notation of the previous paragraph assume either

(a) $n = 1$

(b) $n = 3$ and $v_2(k) \neq 1,2$

or (c) $n = 7$ and $v_2(k) \neq 1,2,3$.

Then E_k is an H-space, s is an H-map and the inclusion $H \to E_k$ is an H-map.

<u>Proof</u>. Since π is a fibre map, E_k can be regarded as the weak pullback of π and $\underline{k}$. Now G operates on $S^n = G/H$ in the usual way and this is an operation in the sense of §1. We wish to use Theorem GZ in this case. To verify that the hypotheses of Theorem GZ hold, we apply Remark (c) of §1. For this observe that $S^n = \Sigma S^{n-1}$ and $\underline{k} = \theta_k$. Thus we must show there exists a multiplication on the domain

S^n and on the range S^n with respect to which $\underline{k}: S^n \to S^n$ is an H-map. But a result of Arkowitz-Curjel as formulated in [6] asserts that this latter statement is equivalent to hypothesis (a), (b) or (c). Theorem GZ then gives the conclusions of the Proposition.

<u>Corollary (HSMNT Theorem)</u>. With the notation of the first paragraph of §2 assume that $n = 1,3$ or 7 and that the classification map $\alpha \in \pi_n(B_H)$ of the principal fibration π has finite order d. If $\nu_2(k) = 0$ or if $\nu_2(k) \geq \nu_2(d)$, then E_k is an H-space.

The proof consists of finding an integer N such that $N \equiv k(d)$ and $\underline{N}: S^n \to S^n$ satisfies the hypothesis of the previous proposition. This is just elementary number theory and hence omitted. Thus $E_N \equiv E_k$ is an H-space.

<u>Remarks</u>. (a) This is only half of the HSMNT theorem (we follow the formulation in [5]). The other half, dealing with n odd and k odd, is completely subsumed under Zabrodsky's theorem (§3).

(b) The hypotheses of the Corollary are weaker than the hypotheses of the HSMNT theorem since the condition is only for the prime 2 and not for all primes. However, we know of no example of a fibration for which the weaker hypothesis is satisfied and the stronger hypotheses is not.

3. ZABRODSKY'S THEOREM AND H-MAPS

For completeness we first indicate how Zabrodsky's theorem [11] follows from Theorem GZ by means of localization. Let (Y,μ_Y) be an H-space and G a topological group with action $\eta: G \times Y \to Y$ such that $\eta(1 \times \mu_Y) = \mu_Y(\eta \times 1): G \times Y \times Y \to Y$. Suppose $Y/G = S^n$ and let $g: Y \to S^n$ be the projection onto the orbit space.

<u>Zabrodsky's Theorem</u>. With the notation above, let P be the weak pullback of $g: Y \to S^n$ and $\underline{k}: S^n \to S^n$ (fig. 4). If k is odd and n is odd, then P is an H-space.

$$\begin{array}{ccc} P & \xrightarrow{s} & Y \\ \downarrow r & & \downarrow g \\ S^n & \xrightarrow{\underline{k}} & S^n \end{array}$$

(figure 4)

$$\begin{array}{ccc} P_\ell & \xrightarrow{s_\ell} & Y_\ell \\ \downarrow r_\ell & & \downarrow g_\ell \\ S^n_\ell & \xrightarrow{\underline{k}_\ell} & S^n_\ell \end{array}$$

(figure 5)

<u>Proof</u>. We use standard facts and notation for localization as found for example in [9]. If ℓ denotes the odd primes, then to show

P is an H-space it suffices to show (1) P_2, the localization of P at 2, is an H-space (2) P_ℓ, the localization of P at ℓ, is an H-space (3) the two induced H-structures on P_0, the rationalization of P, obtained from (1) and (2) are compatible. Since k is odd, the localization of $\underline{k}$ at 2, $\underline{k}_2\colon S_2^n \to S_2^n$, is a homotopy equivalence. Thus $s_2\colon P_2 \to Y_2$ is a homotopy equivalence. Since Y is an H-space, so is Y_2, and we use the equivalence s_2 to induce an H-structure on P_2. We shall prove that P_ℓ is an H-space such that s_ℓ is an H-map. Assuming this for the moment, we can establish (1), (2) and (3). For the two H-structures on P_0 are compatible since they both come from the same induced H-structure on Y_0 via localizations of the map s.

Thus we must prove that P_ℓ is an H-space and $s_\ell\colon P_\ell \to Y_\ell$ an H-map. For this we apply the ℓ-localization functor to the weak pullback diagram fig. 4 to obtain the diagram fig. 5 which is also a weak pullback diagram. Since $2 \notin \ell$ and n is odd, it follows that S_ℓ^n is a homotopy-abelian H-space [1]. Furthermore, the operation of Y on $S^n = Y/G$ induces an operation of Y_ℓ on S_ℓ^n. Therefore to apply Theorem GZ to figure 5 and conclude that P_ℓ is an H-space and s_ℓ an H-map, we appeal to Remark (c) of §1. Since $S_\ell^n = \Sigma S_\ell^{n-1}$ and $\underline{k}_\ell = \theta_k\colon S_\ell^n \to S_\ell^n$, it suffices to show θ_k is an H-map. But this follows since the set $[S_\ell^n \times S_\ell^n, S_\ell^n] \equiv [S^n \times S^n, S_\ell^n]$ with operation inherited from the H-space S_ℓ^n is an abelian group. This completes the proof.

Our final result deals with the use of Theorem GZ to obtain information on H-maps between some of the new H-spaces. For definiteness we concentrate on one of the two main situations to which the HSMNT theorem applies. This concerns the principal fibration $S^3 \to Sp(2) \xrightarrow{\pi} S^7$ with classification map of order 12. For $\underline{k}\colon S^7 \to S^7$ and $0 \leq k \leq 11$, the total space E_k of the principal fibration induced from π via $\underline{k}$ is an H-space if and only if $k \not\equiv 2, 6, 10$ [4,7,10]. We call such an H-space a <u>Hilton-Roitberg-Stasheff H-space</u>. If E_r and E_s are two Hilton-Roitberg-Stasheff H-spaces then a map $\lambda\colon E_r \to E_s$ is said to have <u>base degree</u> $\underline{k}$ if the two squares in the following diagram commute

$$\begin{array}{ccccc} S^3 & \longrightarrow & E_r & \longrightarrow & S^7 \\ \downarrow{\scriptstyle 1} & & \downarrow{\scriptstyle \lambda} & & \downarrow{\scriptstyle \underline{k}} \\ S^3 & \longrightarrow & E_s & \longrightarrow & S^7 \end{array}$$

where the horizontal sequences are fibrations.

Proposition. Let E_s be a Hilton-Roitberg-Stasheff H-space and k an integer with $\nu_2(k) \neq 1,2,3$. Then there exists a Hilton-Roitberg-Stasheff H-space E_r and an H-map $\lambda: E_r \to E_s$ of base degree k.

Proof. By Remark (b) of §1 there is an operation of E_s on S^7 relative to the fibre map $g: E_s \to S^7$. Since $\nu_2(k) \neq 1,2,3$, $\underline{k}: S^7 \to S^7$ is an H-map. By Remark (c) of §1 and Theorem GZ the (weak) pullback P of g and $\underline{k}$ is an H-space and the projection $\lambda: P \to E_s$ is an H-map. But $P \simeq E_r$ for some r, $0 \leq r \leq 11$; in particular, choose $r \equiv ks$ (12). Thus $\lambda: E_r \to E_s$ is an H-map of base degree k.

Concluding Remarks. The proposition shows that there are many different H-maps between Hilton-Roitberg-Stasheff H-spaces. The proof also shows how to find r. Thus for particular r,s and k one can often determine whether there is an H-map $E_r \to E_s$ of base degree k.

We also remark on the other basic situation, due to Curtis and Mislin [2], to which the HSMNT theorem applies. Here one starts with the principal fibration $SU(3) \to SU(4) \xrightarrow{\nu} S^7$ with classification map of order 6 and denotes by X_k the fibre space induced from ν via $\underline{k}: S^7 \to S^7$, $0 \leq k \leq 5$. Then all the X_k are H-spaces and there is an analogue to the proposition for the Curtis-Mislin H-spaces X_k.

REFERENCES

[1] Adams, J.F., The sphere, considered as an H-space mod p, Quart. J. Math. Oxford (2) 12(1961), 52-60.

[2] Curtis, M. and Mislin, G., H-spaces which are bundles over S^7, J. of Pure and Applied Algebra 1 (1971), 27-40.

[3] Harrison, J. and Stasheff, J., Families of H-spaces, Quart. J. Math. Oxford (2) 22 (1971), 347-351.

[4] Hilton, P.J. and Roitberg, J., On principal S^3-bundles over spheres, Ann. of Math. 90 (1969), 91-107.

[5] Mimura, M., Nishida, G., and Toda, H., Localizations of CW-complexes and its applications, J. Math. Soc. Japan 23 (1971), 593-624.

[6] Sigrist, F., H-maps between spheres, H-spaces, Lecture Notes in Mathematics, Springer-Verlag, No. 196 (1970), 39-41.

[7] Stasheff, J., Manifolds of the homotopy type of (non-Lie) groups, Bull. Amer. Math. Soc. 75 (1969), 998-1000.

[8] Stasheff, J., Families of finite H-complexes-revisited, H-spaces, Lecture Notes in Mathematics, Springer-Verlag, No. 196 (1970), 1-4.

[9] Sullivan, D., Geometric Topology, Part 1, rev. ed., M.I.T., Cambridge, Mass., 1971.

[10] Zabrodsky, A., On sphere extensions of classical Lie groups, Proc. of Symposia in Pure Math., Vol. 22, Amer. Math. Soc., Providence, R.I., 1971.

[11] ___________ , On the construction of new finite CW H-spaces, Invent. Math. 16 (1972), 260-266.

A FUNCTOR WHICH LOCALIZES THE HIGHER HOMOTOPY GROUPS OF AN ARBITRARY C. W. COMPLEX

Martin Bendersky, University of Washington

§1. THE SEMI-LOCALIZATION

Let $\mathcal{C}$ be the category of pointed spaces with homotopy type of a C. W. complex.

Let $K = \bigcup Y_f$, $f \in \Omega^2 X, Y_f = S^2_P$ for all f. (P is a fixed set of primes.) $L_P(X)$ is defined to be $X \bigcup K/\sim$ where $k \sim f(k)$ if $k \in S^2 \subset Y_f$. Let ε'_P be the inclusion. Clearly $L_P: \mathcal{C} \to \mathcal{C}$ is a functor.

For $X \in \mathcal{C}$, set $M = \{f: \Sigma^k M(Z/m,1) \to X,\ k \geq 1,\ m$ invertible in $Z_P\}$ where $M(Z/m,1)$ is a standard Moore space.

We define $R^1_P(X)$ to be the space obtained from X by attaching cones on all maps $f \in M$.

$$R^{i+1}{}_P(X) = R^1_P(R^i_P(X)); \quad R_P(X) = \bigcup R^i_P(X).$$

Let $\bar{\varepsilon}_P: X \to R_P(X)$ be the inclusion. Notice that $R_P(X)$ differs from the functor introduced in Anderson [1] in that we do not attach cones on unsuspended Moore spaces.

Definition 1.1. The semi-localization $X_{\bar{P}}$ is defined to be $R_P \circ L_P(X)$ a natural transformation from the identity to the semi-localization is given by $\varepsilon_P = \bar{\varepsilon}_P \circ \varepsilon'_P$.

Clearly the semi-localization is a functor on $\mathcal{C}$.

In the sequel we shall usually drop the subscript $\bar{P}$.

Theorem 1.2. The semi-localization enjoys the following properties.

i) $\pi_1(\varepsilon)$ is a natural isomorphism

ii) $\pi_*(\varepsilon)$ is the localization for $* \geq 2$

iii) $H_*(\varepsilon)$ is a P-bijection

iv) $\widetilde{(X_{\bar{P}})} = (\tilde{X})_{\bar{P}}$ where $\sim$ denotes the universal cover

v) In the diagram

$$\begin{array}{ccc} X & \longrightarrow & X_{\bar{P}} \\ & \searrow & \downarrow f \\ & & Y \end{array}$$

there is an f making the diagram commute if Y is semi-local (i.e. $\pi_*(Y)$ is local for $* \geq 2$). Furthermore, if Y is nilpotnet, f is unique, up to homotopy.

As a consequence if $k: X \to Y$ is any map satisfying (i) and (ii) then Y must be of the same homotopy type as $X_{\bar{P}}$.

The proof of 1.2 may be found in Bendersky [2].

The homology of $X_{\bar{P}}$ is easy to calculate

(1.3)

$H_*(X_{\bar{P}}; Z/p) = H_*(X; Z/p)$ if $p \in P$

$H_*(X_{\bar{P}}; Z/p') = H_*(\pi_1(X); Z/p')$ if p' is invertible in Z_P.

As a corollary we have

(1.4)

If $\pi_1(X)$ is a nilpotent P-local group $H_*(\mathcal{E})$ localizes homology.

In this case we simply write X_P for $X_{\bar{P}}$.

For example one may localize RP^{2n}, $BF(k)$ and $BO(k)$ at 2.

§2. APPLICATION

A.

One may use the semi-localization in conjunction with the functors of Hilton-Mislin-Roitberg [7] or Bousfield-Kan [9] to localize a larger class of spaces than nilpotent ones.

<u>Definition 2.1.</u> X is nilpotent mod C_P if for the descending central series of $\pi_n(X)$ (Bousfield-Kan [ ]) Γ_r is tosion prime to P.

One may localize such spaces by first semi-localizing. $X_{\bar{P}}$ is now nilpotent hence $X \to (X_{\bar{P}})_P$ localizes.

B.

As with localization one may build up a space from $\{X_{\bar{P}_i}\}$ for a finite partition of primes $\{P_i\}$. To be precise we have

<u>Proposition 2.2.</u>

The pullback of the diagram

$$\begin{array}{ccc} & & X_{\bar{P}_2} \\ & & \downarrow \\ X_{\bar{P}_1} & \longrightarrow & X_{\overline{P_1 \cap P_2}} \end{array}$$

is $X_{\overline{P_1 \cup P_2}}$.

Note that the semi-localizations in 2.2 are only of the same homotopy type as that constructed in §1.

C.

In the situation of a mapping space X^W, (where W is a finite, path connected C.W. complex), the component of any map which admits a lift to $\tilde{X}$ is nilpotent. Using the semi-localization we may write the localization of X^W as a mapping space.

<u>Proposition 2.3.</u>

$$(X^W)_f \xrightarrow{\varepsilon} (X^W_{\bar{P}})_{\varepsilon \circ f}$$

localizes <u>all</u> homotopy groups if $\pi_1(f) = 0$.

One may then partially generalize the pull back theorem of Hilton Mislin and Roitberg [7]. For example, if a map g, semi-localizes to the constant map at all primes then, $g \sim *$.

From this one may construct candidates for non-cancellation using projective bundles over spheres.

Details, and further applications will appear in Goodisman [4].

§3. FIBRE WISE LOCALIZATION

Let $F(q)$ be the monoid of pointed degree ± 1 maps of S^q to S^q.

There is also the monoid $F(q)_P$ of pointed degree ± 1 maps of S^q_P to S^q_P. Using a localization which is functional, such as in Bousfield-Kan [9], one may define a map loc: $F(q) \to F(q)_P$. The map on classifying spaces is seen to be the semi localization (up to homotopy).

Let $\alpha: E \to B$ be a fibration with fibre S^n, $n > \dim B$. One may consider the S^n_P fibration classified by $\alpha_{\bar{P}}: B \to BF_{\bar{P}}$, and the fibration $\bar{\alpha}_P$ obtained by fibre wise localization (Sullivan [13]).

<u>Proposition 3.1</u>

$$\alpha_{\bar{P}} = \bar{\alpha}_P .$$

Using 3.1 one may prove a generalization of a theorem of Glover and Mislin [3].

<u>Theorem 3.2.</u> Let M and N be connected, smooth n-manifolds. Suppose $M_{\bar{2}} \sim N_{\bar{2}}$, then with k odd $> [\frac{n+1}{2}]$

$$M \subseteq R^{n+k} \Leftrightarrow N \subseteq R^{n+k} .$$

There are two basic results whose generalizations require the semi-localization.

The first is a decomposition of $BF(q)$ (q odd) into its local components.

<u>Proposition 3.3.</u> Let X be the n+1 skeleton of $BF(q)$, q odd, $> [\frac{n+1}{2}]$, then for dimension $Y \leq n$, there is a bijection

$$[Y, X] \sim \prod_{p>2} [Y, BSF(q)_p] \times [Y, BF(q)_2]$$

Proof: Consider the map

$$g_p: X \to BF(q) \to BF \to BF_p \xrightarrow{\simeq} BSF_p \quad .$$

Since p, and q are odd, with $q > [\frac{n+1}{2}]$ there is a lift of g_p to a map $f_p: X \to BSF(q)_p$. The map

$$\prod_{p>2} f_p \times \mathcal{E}_2: X \to \prod_{p>2} BSF(q)_p \times BF(q)_2$$

induces an isomorphism in π_* for $* \leq n$ and a surjection for $* = n+1$, and 3.3 follows.

The other result we shall need is the following:

Proposition 3.4. Let M and N be closed connected smooth manifolds and $\phi: N_{\bar{2}} \to M_{\bar{2}}$ a homotopy equivalence. Then $\phi^* J\nu(M)_{\bar{2}} = J\nu(N)_{\bar{2}}$ in $[N_{\bar{2}}, BF_2] = [N, BF_2]$.

Granting 3.4 the proof of 3.2 basically follows Glover and Mislin. Let $J\nu(N): N \to BF$ be the stable normal sphere bundle of N. Then for p odd $J\nu(N)_p$ lifts to a map $N \to BSF(k)_p$, since $k > [\frac{n+1}{2}]$, (see the proof of 3.3).

By 3.4 we have a lift of $J\nu(N)_2: N \to BF(k)_2$. Hence by 3.3 we obtain a lift of $J\nu(N)$ to $N \to BF(k)$. Since $k > [\frac{n}{2}]$ this implies $\nu(N)$ lifts to a map $N \to BO(k)$ (Sutherland [14]). It follows from Hirsch [8] that N immerses in R^{n+k}.

Proof of 3.4. Suppose $\phi^*(J\nu(M)_{\bar{2}}) = w \neq J\nu(N)_{\bar{2}}$. Then with Q = {odd primes} we define an element $\hat{\theta} \in [N, BF_2] \times [N, BF_{\bar{Q}}]$ as follows. $\hat{\theta}_2 = w$, $\hat{\theta}_{\bar{Q}} = J\nu(N)_{\bar{Q}}$.

By the Wu formula and the equivalence ϕ, it follows that $\hat{\theta}_2$ and $\hat{\theta}_{\bar{Q}}$ both map to the same element in $[N, BF_{\bar{0}}] = [N, K(Z/2,1)]$. Hence by 2.2 there is a $\theta \in [N, BF]$ lifting $\hat{\theta}$. θ is not $J(\nu(N))$. Following Glover and Mislin, we obtain a contradiction by showing that $(N)^{\theta}$, the Thom space of θ is S-reducible. A theorem of Spivak

[12] would imply $\theta = J(\nu(N))$. By the pullback theorem of Hilton-Mislin-Roitberg [7] it suffices to show that $(N)^{\theta}_{p}$ is S-reducible at p for each prime p. We first consider the case of an odd prime. For a spherical fibration $\alpha: E \to B$ $\bar{\alpha}_p: E' \to B$ denotes the fibre-wise localization. $(N)^{\bar{\alpha}_p}$ is defined to be the mapping cone of $\bar{\alpha}_p$.

Lemma 3.5.

$$(N)^{\alpha}_{p} = (N)^{\bar{\alpha}_p}$$

Proof: If α is fibre homotopically trivial, then $(N)^{\alpha} \to (N)^{\bar{\alpha}_p}$ is, up to homotopy, the natural map $S^k(N/\phi) \to S^k_p(N/\phi)$ (k large), which is the localization. Give N some triangulation. Then 3.5 is true over each simplex. An argument using the relative Meyer-Vietoris sequence, and the p-local 5 lemma proves 3.5 in general.

We now have for p odd,

$$\begin{aligned}(N)^{\theta}_{p} &= (N)^{\bar{\theta}_p} \\ &= (N)^{\theta_{\bar{p}}} \\ &= (N)^{J\nu(N)_{\bar{p}}} \\ &= (N)^{J\nu(N)}_{p}\end{aligned}$$

which is S-reducible at p.

For $p = 2$, we have for any fibration $\alpha: S^n \to E \xrightarrow{t} N$ with $n > 1$ a quasi-fibration.

$$\alpha(p): S^n_p \to E_{\bar{p}} \xrightarrow{t_{\bar{p}}} N_{\bar{p}}$$

We define $(N_{\bar{p}})^{\alpha(p)}$ to be the mapping cone of $t_{\bar{p}}$.

Lemma 3.6. The natural map $\eta: (N)^{\theta} \to (N_{\bar{p}})^{\theta(p)}$ is the localization.

Proof. To show that $(N_{\bar{p}})^{\theta(p)}$ is local we show that $\tilde{H}_*((N_{\bar{p}})^{\theta(p)};Z/m) = 0$ for m a prime $\neq$ p. There is an exact sequence

$$\to H_*(E_{\bar{p}};\ Z/m) \to H_*(N_{\bar{p}};\ Z/m) \to H_*((N_{\bar{p}})^{\theta(p)};\ Z/m) \to$$

By 1.3 $(t_{\bar{p}})_*$ is an isomorphism. As for the p-bijectivity of $H_*(\eta)$ we apply 1.2 (iii) and the local 5 lemma to the diagram

$$\begin{array}{ccccccc} \to H_*(E) & \to & H_*(N) & \to & H_*((N)^{\theta}) & \to \\ \downarrow & & \downarrow & & \downarrow & \\ \to H_*(E_{\bar{p}}) & \to & H_*(N_{\bar{p}}) & \to & H_*((N_{\bar{p}})^{\theta(p)}) & \to \end{array}$$

We note that if one first fibrewise localizes θ, then applies the semi-localization to the resulting total space and the base, the fibration obtained is $\theta(p)$. There is the mapping of fibrations

$$\theta \to \theta_{\bar{2}} = w \to w(2) \overset{\phi}{\to} J\nu(M)(2) \leftarrow J\nu(M)$$

From which it follows that

$$(N)_2 = (M)_2^{J\nu(M)}$$

and $(N)^{\theta}$ is S-reducible at 2. This completes the proof of 3.4.

§4. REMARKS ON EMBEDDINGS

A theorem similar to 3.4 for embeddings was proven by R. Rigdon [11]. The following propostion, first proven by E. Rees is a 2-local version of a theorem of Haefliger and Hirsch [6], [10]

Proposition 4.1. Let M be a smooth, compact, connected n-manifold with out boundary. Suppose $\tilde{H}_i(M)$ is odd torsion for $i \leq k < n/2$, k even. Then $M \subset R^{2n-k}$.

Proof. Let $M^* = (M \times M - \Delta)/Z_2$ where Δ denotes the diagonal, and Z_2 acts on $M \times M - \Delta$ by switching factors. The methods of Haefiger-Hirsch [6] show that $H^i(M^*;\ Z)$ is odd torsion for $i \geq 2n - k$. The Z_2 fibration $M \times M - \Delta \to M^*$ has a classifying map $M^* \to RP^{\infty}$. By a theorem of Haefliger [5] 4.1 will be proven if we can find a lift h in the diagram

$$\begin{array}{ccc} & & RP^{2n-k-1} \\ & \overset{h}{\nearrow} & \downarrow \pi \\ M^* & \longrightarrow & RP^{\infty} \end{array}$$

Since k is even the fibration π is simple. We shall need the following.

<u>Lemma 4.2.</u> Let k be an integer, t odd. Then there is a map $\bar{k}: RP^t \to RP^t$ inducing a map of degree k on the universal covers. Furthermore, if k is odd, the diagram

$$\begin{array}{ccc} RP^t & \xrightarrow{\bar{k}} & RP^t \\ & \searrow \quad RP^{\infty} \quad \swarrow & \end{array}$$

commutes.

The obstructions to finding h in 4.2 lie in $H^{i+1}(M^*;{}'\pi_i(S^{2n-k-1}))$. (For an abelian group G, $'G$ denotes $G/(2\text{-primary torsion})$). Suppose we have obtained a lift h_r, of M^* to the rth stage of the Postnikov resolution of Υ. Let the obstruction $\mathcal{O}$ have order d. Then d is odd. Let $\bar{d}$ be the map given by 4.2

There is an induced map of Postnikov resolutions $\{\bar{d}_r\}$. $\bar{d}_r \circ h_r$ is a new lift of M^*, and the new obstruction is the image of $\mathcal{O}$ via the map $H^{r+1}(M^*;{}'\pi_r(S^{2n-k-1})) \to H^{r+1}(M^*;{}'\pi_r(S^{2n-k-1}))$ induced by the coefficient map

$$\bar{d}_{\#}: {}'\pi_r(S^{2n-k-1}) \to {}'\pi_r(S^{2n-k-1}) \quad .$$

However since k is even, $\bar{d}_{\#}$ is multiplication by d, and $\bar{d}_r \circ h_r$ lifts. Continuing in this way we obtain the desired map h, proving 4.1.

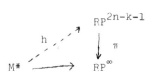

References

1. D. W. Anderson, Localizing C.W. complexes, Ill. Journal of Math. 16 (1972), 519-525.

2. M. Bendersky, Semi localization and a theorem of Glover and Mislin, Univ. of Washington, preprint.

3. H. H. Glover and G. Mislin, Immersion in the metastable range and 2-localization, ETH Zurich, February 1973.

4. L. Goodisman, Ph.D. Thesis, Univ. of Washington.

5. A. Haefliger, Prolongements differentiables dans le domaine, stable, Comm. Math. Helv., 37 (1962), 155-176.

6. A. Haefliger and M. W. Hirsch, Immersions in the stable range, Annals of Math., (1962), 231-241.

7. P. J. Hilton, G. Mislin and J. Roitberg, Homotopical localization, (to appear).

8. M. W. Hirsch, Immersion of manifolds, Trans. Amer. Math. Soc., 93 (1959), 242-276.

9. A. K. Bousfield and D. M. Kan, Limits, completions and localization, Lecture Notes in Math., 304, Springer-Verlag, (1972).

10. E. Rees, Embedding odd torsion manifolds, Bulletin London Math. Soc., 3 (1971), 356-362.

11. R. Rigdon, p-equivalences and embeddings of manifolds, (to appear).

12. M. Spivak, Spaces satisfying Poincaré duality, Topology 6 (1967), 77-101.

13. D. Sullivan, Geometric topology part I, localization, periodicity, and Galois symmetry, MIT, June (1970), (mimeographed notes).

14. W. A. Sutherland, Fibre homotopy equivalence and vector fields, Proc. London Math. Soc., (3), 15 (1965), 543-556.

HOMOLOGICAL LOCALIZATIONS OF SPACES, GROUPS, AND Π-MODULES

A.K. Bousfield[1]

University of Illinois at Chicago Circle

§0. Introduction

Let h_* be a generalized homology theory defined on arbitrary CW complexes and satisfying the limit axiom. In §1 we will show that h_* determines a localization functor on the pointed homotopy category, Ho, of CW complexes; and we will discuss the behavior of this localization in detail when h_* is a connective theory. This discussion will involve some new localization functors for groups and π-modules which we will explain in §2. In §3 we will outline a proof of our main existence theorem (1.1) for localizations and will mention a generalization of it. Full details will appear in [2], [3], and elsewhere.

§1. h_*-localizations of spaces

A space $X \in \underline{Ho}$ will be called h_*-*local* if each h_*-equivalence $A \longrightarrow B \in \underline{Ho}$ induces a bijection $[B,X] \approx [A,X]$. We define an h_*-*localization* of $X \in \underline{Ho}$ to be an h_*-equivalence $X \longrightarrow \hat{X}_{h_*} \in \underline{Ho}$ such that $\hat{X}_{h_*}$ is h_*-local. Such an h_*-localization of X is clearly unique up to equivalence and our main theorem asserts its existence.

Theorem 1.1 For each $X \in \underline{Ho}$ *there exists an* h_**-localization* $X \longrightarrow \hat{X}_{h_*}$.

We remark that Adams [1] and Deleanu [5] have an interesting cate-

[1]Research supported in part by NSF grant GP-3894.

gorical approach to theorems of this sort.

In view of the above theorem, there is an obvious h_*-localization functor $(\ )^{\wedge}_{h_*}: \underline{Ho} \longrightarrow \underline{Ho}$ with $(X^{\wedge}_{h_*})^{\wedge}_{h_*} \approx X^{\wedge}_{h_*}$. Roughly speaking, this functor selects a canonical homotopy type within each h_*-homology type. It is also worth noting that $X \longrightarrow X^{\wedge}_{h_*}$ has a universal property: it is the terminal example of an h_*-equivalence going out of X.

We now turn to the problem of "computing" h_*-localizations when h_* is connective, i.e., $h_i(\text{point}) = 0$ for i sufficiently small. We can actually confine our attention to very few homology theories because:

Proposition 1.2 *If h_* is a connective homology theory, then h_* has the same equivalences* (and thus gives the same localizations) *as $H_*(\ ;A)$ where either $A = Z[J^{-1}] \subset Q$ or $A = \bigoplus_{p \in J} Z/pZ$ for some set J of primes.*

Using previous results (see Bousfield-Kan [4]), it is now easy to determine

1.3 The h_*-localization of a nilpotent space for connective h_*. Let $X \in \underline{Ho}$ be a connected nilpotent space, i.e., $\pi_1 X$ is a nilpotent group which is acting nilpotently on $\pi_n X$ for $n \geq 2$. For instance, X could be a simply connected or simple space.

Case (i) Let $h_* = H_*(\ ;Z[J^{-1}])$. Then $X \longrightarrow X^{\wedge}_{h_*}$ is the usual $Z[J^{-1}]$-localization with $\pi_i X^{\wedge}_{h_*} \approx Z[J^{-1}] \otimes \pi_i X$ for $i \leq 1$, where $Z[J^{-1}] \otimes (\ )$ is the Malcev-Lazard completion for nilpotent groups (see Bousfield-Kan [4], p. 128).

Case (ii) Let $h_* = H_*(\ ;Z/pZ)$. Then there is a splittable short exact sequence

$$0 \longrightarrow \text{Ext}(Z_{p^\infty}, \pi_n X) \longrightarrow \pi_n X_{h_*} \longrightarrow \text{Hom}(Z_{p^\infty}, \pi_{n-1} X) \longrightarrow 0$$

involving the Ext and Hom completions for nilpotent groups (see Bousfield-Kan [4], p. 165). When the groups $\pi_* X$ are finitely generated, $\pi_* X\hat{}_{h_*}$ is just the p-profinite completion of $\pi_* X$. This case has been studied by Sullivan [10].

<u>Case (iii)</u> Let $h_* = H_*(\ ; \oplus_{p\in J} Z/pZ)$. Then

$$X\hat{}_{h_*} \simeq \prod_{p\in J} X\hat{}_{H_*(\ ;Z/pZ)}$$

When the groups $\pi_* X$ are finitely generated and J consists of all primes, $\pi_* X\hat{}_{h_*}$ is just the profinite completion of $\pi_* X$. This case has been studied by Sullivan [10].

For non-nilpotent spaces, no simple formula can describe the effect of h_*-localization upon homotopy groups.

<u>Example 1.4</u> If $h_* = H_*(\ ;Z)$, then

$$\pi_i K(\Sigma_\infty, 1)\hat{}_{h_*} \approx \pi_i \Omega^\infty S^\infty$$

for $i \geq 1$ where Σ_∞ denotes the infinite symmetric group (see [7]).

<u>Example 1.5</u> If $h_* = H_*(\ ;Z)$, then

$$\pi_i (RP^2)\hat{}_{h_*} = \begin{cases} Z/2Z & \text{if } i = 1 \\ \underline{Z}_2 & \text{if } i = 2 \\ \Gamma(\underline{Z}_2) & \text{if } i = 3 \end{cases}$$

where $\underline{Z}_2$ denotes the 2-adic integers and Γ is J.H.C. Whitehead's homogeneous quadratic functor. Moreover, for $i \geq 4$ $\pi_i (RP^2)\hat{}_{h_*}$ is the sum of the 2-torsion in $\pi_i RP^2$ with a (sometimes huge) vector space over Q.

Although the effect of h_*-localization on homotopy groups is often radical, we can nevertheless detect h_*-local spaces by looking at their homotopy groups. Using the notion of an "HR-local" group (see 2.1) and of an "HZ-local" π-module (see 2.7), we have

Theorem 1.6 Let $h_* = H_*(\ ;R)$ where $R = Z[J^{-1}]$ or $R = Z/pZ$. A connected space $X \in Ho$ is h_*-local if and only if the groups $\pi_i X$ are HR-local for $i \geq 1$ and the $\pi_1 X$-modules $\pi_i X$ are HZ-local for $i \geq 2$.

Using this theorem and some methods of Dror, we can prove the following Whitehead-like result.

Proposition 1.7 Let $h_* = H_*(\ ;R)$ where $R = Z[J^{-1}]$ or $R = Z/pZ$, and let $f:X \longrightarrow Y \in Ho$. Suppose for some $n \geq 1$ that $f_*:h_i X \longrightarrow h_i Y$ is iso for $i < n$ and epi for $i = n$. Then $f_*:\pi_i \hat{X}_{h_*} \longrightarrow \pi_i \hat{Y}_{h_*}$ is also iso for $i < n$ and epi for $i = n$.

Theorem 1.6 yields other dividends. For instance, there is a (somewhat delicate) step by step procedure for constructing $H_*(\ ;R)$-localizations of CW complexes by attaching cells so as to successively localize homotopy groups.

§2. Homological localizations of groups and π-modules

In our discussion of the h_*-localization for spaces, we have spoken of "HR-local" groups and "HZ-local" π-modules. We will now develop the algebraic localization theories underlying these notions.

2.1 The HR-localization for groups Let R be a prime field $(R = Z/pZ)$ or a subring of the rationals $(R = Z[J^{-1}])$. A group homomorphism $f:A \longrightarrow B$ will be called an HR-equivalence if $f_*:H_i(A;R) \longrightarrow H_i(B;R)$ is iso for $i = 1$ and epi for $i = 2$. A group G will be called HR-local if each HR-equivalence $A \longrightarrow B$ induces a bijection

$$\mathrm{Hom}_{(\mathrm{groups})}(B,G) \approx \mathrm{Hom}_{(\mathrm{groups})}(A,G).$$

We define an HR-localization of a group G to be an HR-equivalence

$G \longrightarrow \hat{G}_R$ such that $\hat{G}_R$ is HR-local. This is clearly unique up to equivalence, and

<u>Theorem 2.2</u> <u>For each group G, there exists an HR-localization</u> $G \longrightarrow \hat{G}_R$.

This result can be proved topologically using

<u>Proposition 2.3</u> <u>If G is a group and</u> $h_* = H_*(\ ;R)$, <u>then</u> $\pi_1 K(G,1)\hat{}_{h_*} \approx \hat{G}_{h_*}$.

We remark that $G \longrightarrow \hat{G}_{h_*}$ is functorial and has a universal property: it is the terminal example of an HR-equivalence going out of G.

We will now give examples of HR-localizations. For a group G let $G = \Gamma_1 G \supset \Gamma_2 G \supset \ldots$ denote the lower central series.

<u>Example 2.4</u> Let $R = Z[J^{-1}]$ and suppose $R \otimes (\Gamma_n G/\Gamma_{n+1} G) = 0$ for some $n \geq 1$ (This is automatic if G is finite, nilpotent, or perfect). Then $\hat{G}_R \approx R \otimes (G/\Gamma_n G)$, i.e., $\hat{G}_R$ is the Malcev-Lazard completion of $G/\Gamma_n G$.

<u>Example 2.6</u> If G is a group such that $G/\Gamma_2 G$ is finitely generated, then there is a natural epimorphism

$$\hat{G}_Z \longrightarrow \lim_{\substack{\leftarrow \\ n}} G/\Gamma_n G.$$

We do not know whether this is an isomorphism when G is free.

We now turn to

<u>2.7 The HZ-localization for π-modules</u> Let π be a fixed group. A π-module homomorphism $f:A \longrightarrow B$ will be called an HZ-<u>equivalence</u> if $f_*:H_i(\pi;A) \longrightarrow H_i(\pi;B)$ is iso for $i = 0$ and epi for $i = 1$. A π-module M will be called HZ-<u>local</u> if each HZ-equivalence

$A \longrightarrow B$ induces a bijection

$$\mathrm{Hom}_{(\pi\text{-modules})}(B,M) \approx \mathrm{Hom}_{(\pi\text{-modules})}(A,M).$$

We define an HZ-localization of a π-module M to be an HZ-equivalence $M \longrightarrow \hat{M}$ such that $\hat{M}$ is HZ-local. This is clearly unique up to equivalence, and

Theorem 2.8 *For each π-module M, there exists an HZ-localization $M \longrightarrow \hat{M}$.*

The HZ-localization $M \longrightarrow \hat{M}$ is functorial on π-modules and has a universal property: it is the terminal example of an HZ-equivalence going out of M.

We will now give examples of HZ-localizations. Let $I \subset Z\pi$ be the augmentation ideal.

Example 2.9 If M is a π-module such that $I^n M = I^{n+1} M$ for some n, then $\hat{M} \approx M/I^n M$. In particular, any nilpotent π-module is HZ-local.

Example 2.10 Let π be a finitely generated nilpotent group and let M be a finitely generated π-module. Then

$$\hat{M} \approx \varprojlim_n M/I^n M$$

Example 2.11 If $\pi = Z/2Z$ acts on an abelian group M by negation, then

$$\hat{M} \approx \mathrm{Ext}(Z_{2^\infty},M).$$

§3. Proof and generalization of Theorem 1.1 To prove 1.1 we will construct an h_*-localization functor on the category, $\underline{S}$, of simplicial sets which induces the desired h_*-localization functor on $\underline{Ho}$

(Full details will appear in [2]). As a byproduct of this proof we introduce a version of simplicial homotopy theory in which h_*-equivalences play the role of weak homotopy equivalences. We conclude with a generalization of 1.1 which yields h_*-localizations of spaces over a fixed space.

Let h_* be as in §0.

Definition 3.1 $K \in \underline{S}$ is an h_*-Kan complex if it has the extension property for pairs $L \subset M \in \underline{S}$ with $h_*(M,L) = 0$, i.e., any map $L \longrightarrow K \in \underline{S}$ can be extended over M.

The h_*-Kan complexes are automatically Kan complexes. They are useful to us because

Lemma 3.2 *If K is a pointed h_*-Kan complex, then $|K| \in \underline{Ho}$ is h_*-local.*

Our next lemma gives a "small" criterion for detecting h_*-Kan complexes. Let c be a fixed infinite cardinal number not less than the cardinality of $h_*(\text{point})$.

Lemma 3.3 *$K \in \underline{S}$ is an h_*-Kan complex if and only if it has the extension property for pairs $L \subset M \in \underline{S}$ such that M has at most c simplices and $h_*(M,L) = 0$.*

We can now show that there are "enough" h_*-Kan complexes in $\underline{S}$.

Lemma 3.4 *For each $K \in \underline{S}$ there exists an h_*-equivalence $K \longrightarrow \hat{K} \in \underline{S}$ such that $\hat{K}$ is an h_*-Kan complex.*

Sketch of proof. We construct $\hat{K}$ as the direct limit of a sequence in $\underline{S}$

$$K = K(0) \longrightarrow K(1) \longrightarrow \cdots \longrightarrow K(j) \longrightarrow \cdots$$

indexed by the section of the first ordinal of cardinality greater than c. We form $K(j+1)$ from $K(j)$ by "attaching h_*-acyclic cells corresponding to the extension problems in 3.3," and when j is a limit ordinal we let

$$K(j) = \lim_{\substack{\to \\ i<j}} K(i).$$

Then $K \longrightarrow \hat{K}$ is clearly an h_*-equivalence, and $\hat{K}$ is an h_*-Kan complex by 3.3.

Using the equivalence of simplicial and topological homotopy theory, one can now deduce 1.1 from

3.5 The existence of an h_*-localization functor for simplicial sets

Our above construction of $K \longrightarrow \hat{K} \in \underline{S}$ is functorial and may be viewed as an h_*-localization in $\underline{S}$. Indeed, 3.2 implies that $|K| \longrightarrow |\hat{K}|$ is an h_*-localization in $\underline{Ho}$ when K is pointed. We remark that our h_*-localization in S can be used to construct fibrewise h_*-localization for fibrations.

Our simplicial methods will also give

3.6 A Quillen model category In his work on homotopical algebra [8], Quillen formulated axioms for a closed model category and developed elementary homotopy theory in that framework. We can show that Quillen's axioms hold in $\underline{S}$ when one interprets a "weak equivalence" as an h_*-equivalence, a "cofibration" as an injection, and a "fibration" as a map having the right lifting property with respect to injections which are h_*-equivalences.

Using Quillen's factorization axiom, one can easily prove a generalization of 1.1.

Theorem 3.7 For each map f in $\underline{Ho}$ there exists a terminal example among the factorizations $f = ji$ in $\underline{Ho}$ with i an h_*-equivalence.

The canonical factorization provided by this theorem can be viewed as an h_*-localization of a space (the source of f) over a fixed space (the target of f). The map $X \longrightarrow * \in Ho$ has canonical factorization $X \longrightarrow \hat{X}_{h_*} \longrightarrow * \in Ho$ which gives our usual h_*-localization, and the map $* \longrightarrow X \in Ho$ has canonical factorization $* \longrightarrow A_{h_*}X \longrightarrow X \in Ho$ which gives Dror's acyclic functor [6] when $h_* = H_*(\ ;Z)$.

References

[1] J.F. Adams, Mathematical lectures, University of Chicago, Chicago, 1973.

[2] A.K. Bousfield, The localization of spaces with respect to homology (to appear).

[3] A.K. Bousfield, Types of acyclicity (to appear).

[4] A.K. Bousfield and D.M. Kan, Homotopy limits, completions and localizations, Lecture Notes in Math. 304, Springer (1967).

[5] A. Deleanu, Existence of the Adams completion for CW complexes (to appear).

[6] E. Dror, Acyclic spaces, Topology 11(1972), 339-348.

[ ] S.B. Priddy, On $\Omega^\infty S^\infty$ and the infinite symmetric group, Proc. Symp. Pure Math. AMS 22(1971), 217-220.

[8] D.G. Quillen, Homotopical algebra, Lecture Notes in Math 43, Springer (1967).

[9] D.G. Quillen, An application of simplicial profinite groups, Comm. Math. Helv. 44(1969), 45-60.

[10] D. Sullivan, Geometric topology, part I, M.I.T. (1970).

NORMALIZERS OF MAXIMAL TORI

Morton Curtis*, Alan Wiederhold, Bruce Williams*

Rice University

1. INTRODUCTION

Our main result is the following.

Theorem 1: Let G_1, G_2 be compact connected semisimple Lie groups and let N_1, N_2 be normalizers of maximal tori in them. Then

$$G_1 \cong G_2 \Leftrightarrow N_1 \cong N_2 .$$

If G is a compact connected semisimple Lie group, T is a maximal torus and N is its normalizer in G, we have

$$0 \to T \to N \to W \to 1$$

where $W = N/T$ is the (finite) Weyl group. In this group extension T is abelian so that W acts on T by inner automorphisms. It is known that the transformation group (T,W) is not always sufficient to distinguish the local isomorphism class of the group--for example they are the same for $Sp(n)$ and $SO(2n+1)$ (See Appendix 1). The standard procedure is to use the adjoint representation which will determine the roots and these (along with the integer lattice) will distinguish nonisomorphic groups. By Theorem 1 we see that this additional information is also present in the isomorphism class of the normalizer. This is not too surprising in view of the close relation of the cohomology of BN and BG (See Appendix 2).

Now the simplest group extension

$$0 \to T \to K \to W \to 1 ,$$

with given T,W and action of W on T, is the semidirect product $T \times_w W$. As a set $T \times_w W$ is just the cartesian product $T \times W$ and

*) Research supported by NSF Grant GP-29A38

its group operation is given by

$$(t_1,w_1)(t_2,w_2) = (t_1t_2^{w_1},w_1w_2)$$

where $t_2^{w_1}$ is the image of t_2 under the action of $w_1 \in W$. The extension K is this semidirect product if and only if the projection map $K \to W$ has a homomorphic cross section.

It seems natural to ask for which simple Lie groups N is the semidirect product, or, as we say, N splits. Theorem 2 is the following table of splittings and nonsplittings. Note that if N for G splits and L is any finite normal subgroup of G, then N for G/L also splits. The converse is false. For, as the table shows, Spin(n) does not split, whereas $SO(n) = \frac{Spin(n)}{center}$ (nodd) splits

Theorem 2: For the simple Lie groups we have.

N splits	N does not split
SU(odd)	SU(even)
SU(even)/center	Sp(n)/center
SO(n)	Spin(n)
G_2	F_4, E_6, E_7, E_8 (or mod centers)

It seems strange that Theorems 1 and 2 do not seem to have been known,* because their proofs require no techniques not known for many years. On the other hand it is not clear that they will be useful in Lie group theory. We were led to this study by our homotopy theoretic approach to Lie group theory. A conjecture as to what the normalizer of a maximal torus in a finite-dimensional H-space should be led to these questions about Lie groups.

Section 2 gives a proof of Theorem 1. In section 3 a criterion for N splitting (Theorem 3) is developed. In section 4 sample

*) we have learned recently that J. Tits had earlier obtained the results of Theorem2 (unpublished). His results are more complete in that he also considered groups obtained by quotienting out proper subgroups of centers.

calculations for one splitting and some nonsplittings are given. It should be clear from these how to complete the table of Theorem 2. For the computationally minded, we record the fact that we have matrix proofs of the splittings and nonsplittings for the classical groups and for G_2 .

2. PROOF OF THEOREM 1

Half of Theorem 1 is trivial. Since any two maximal tori in a compact connected Lie group are conjugate, $G_1 \cong G_2$ implies $N_1 \cong N_2$.

Let G be a compact connected semisimple Lie group and let T be a maximal torus in G . The singular set $S \subset T$ is defined by

$$S = \{t \in T \mid t \text{ belongs at least two maximal tori}\} .$$

Lemma 1: The compact connected semisimple Lie group G is determined up to isomorphism by (T,S) .

Proof: The kernel of the exponential map $\exp: \mathfrak{L}_T \to T$ is called the integer lattice and is denoted by I . The Cartan-Stiefel diagram D in $\mathfrak{L}_T$ is just $\exp^{-1}(S)$. (See [1], page 101). Thus (T,S) determines I and D , and it is well known that I and D determine G up to isomorphism. (See, for example [6], Corollary 6.8, page 270). So Lemma 1 is proved.

Because of Lemma 1 we just need to show that N determines $S \subset T$ to prove Theorem 1. The plan will be to define a set $S(N)$ which is determined by N and then show $S(N) = S$.

Let N be a topological group which is the normalizer of a maximal torus in some compact connected semisimple Lie group. The maximal torus is just the identity component N_0 of N . So we are done if we can give an expression for S in terms of N .

Let $\tilde{N}_0 \xrightarrow{\pi} N_0$ be the universal covering and let $N \longrightarrow N/N_0$ be the quotient map. Since N_0 is abelian, conjugation in N

induces an action of N/N_0 on N_0 and on $\tilde{N}_0$. Of course, $\tilde{N}_0$ is just a real vector space.

Let $R = \{w \in N/N_0 |$ the fixed point set of w is a hyperplane of $\tilde{N}_0 \}$. These are the reflections in N/N_0. Next let $h = \{$hyperplanes $H \subset \tilde{N}_0 |$ H is the fixed point set for some $w \in R\}$. If $w \in R$, $\rho^{-1}(w) \subset N$ and we denote by $K(w)$ the image of $\rho^{-1}(w)$ under the squaring map $x \longmapsto x^2$. Finally let

$$S(N) = \bigcup_{H \in h} \pi(H) \cup \bigcup_{w \in R} K(w) .$$

It is clear from the above that N determines $S(N)$.

Now let G be a compact connected semisimple Lie group and let T be a maximal torus in G. Let $\theta_i : \mathfrak{L}_T \longrightarrow R$ be a root of G and let N_G be the normalizer of T in G. For each integer $n \in \mathbf{Z} \subset R$ we set

$$H_{i,n} = \theta_i^{-1}(n) ,$$

and let w_i be the unique element in the Weyl group $W = N_G/T$ such that $H_{i,0}$ is the fixed point set of w_i. Let $\rho_G : N_G \longrightarrow W$ be the quotient map.

Lemma 2: $\exp(\bigcup_{n \in \mathbf{Z}} H_{i,n}) = \exp H_{i,0} \cup K(w_i)$. (Here $K(w_i)$ is the image of $\rho_G^{-1}(w_i)$ under the squaring map.)

Once Lemma 2 is proved we can finish our proof of Theorem 1 by

Corollary 3: $S = S(N)$.

Proof: Let $\pm\theta_1, \ldots, \pm\theta_k$ be the set of roots of G with respect to T. By [5] (VI 1.5 Théorème 2(iv)) we have that $w \in R$ if and only if $w = w_i$ for some root θ_i. Thus

$$S = \exp(\bigcup_{i,n} H_{i,n}) \text{ , and by Lemma 2}$$

$$S = \bigcup_i \exp H_{i,0} \cup \bigcup_i K(w_i)$$

$$= \bigcup_{H\in h} \pi(H) \cup \bigcup_{w\in R} K(w) = S(N) .$$

The remainder of this section is devoted to a proof of Lemma 2. The ideas needed go back at least as far as Hopf [7].

Let G_i be the identity component of the centralizer of $\exp(\bigcup_{n\in\mathbb{Z}} H_{i,n})$. The following facts about the subgroups G_i are easily checked.

(1) T is also a maximal torus for each G_i

(2) The singular set S_i of G_i is given by

$$S_i = \exp(\bigcup_{n\in\mathbb{Z}} H_{i,n})$$

(3) If τ_i denotes the unique element in $\mathfrak{L}_T$ such that for any $X \in \mathfrak{L}_T$ we have $w_i(X) = X - \vartheta_i(X)\tau_i$, then

$$H_{i,n} = H_{i,0} + \tfrac{n}{2}\tau_i .$$

(4) $\exp(\tau_i) = 0$ (i.e., $\tau_i \in$ the integer lattice I), and

$$\exp(\bigcup_{n\in\mathbf{Z}} H_{i,n}) = \exp H_{i,0} \cup \exp H_{i.1} .$$

(5) If N_i is the normalizer of T in G_i, then

$$N_i = T \cup \rho_G^{-1}(w_i) .$$

(6) If we think of S^3 as quaternions of unit norm, $S^1 = \{\cos\theta + i \sin\theta\}$ is a maximal torus.

For $i = 1,\ldots,k$ there is a homomorphism

$$\beta_i : S^3 \longrightarrow G_i$$

with the following properties:

(a) $\beta_i(S^3)$ and T generate G_i

(b) $\beta_i(S^3) \cap T = \beta_i(S^1)$

(c) Let $\overline{N}$ be the normalizer of S^1 in S^3. Then

$\beta_i(\overline{N}) \subset N_i$, and the single nonidentity component M of $\overline{N}$ goes into $\rho_G^{-1}(w_i)$.

(d) The squaring map sends M to $(-1) \in S^3$ and

$$\beta_i(-1) = \exp(\frac{\tau_i}{2}) \in \exp H_{i,1} .$$

(For this last fact see [4] Chapter III,3.)

Now 6(c) and 6(d) imply that $K(w_i)$ contains at least one element x such that $x^2 \in \exp H_{i,1}$. From (4) we see that to finish the proof of Lemma 2, it suffices to show that $K(w_i) = \exp H_{i,1}$.

If $x \in \rho_G^{-1}(w_i)$ then $x \notin T$, but there exists some maximal torus T' of G_i with $x \in T'$. Then $x^2 \in T \cap T'$ and hence is singular. Thus $K(w_i) \subset S_i = \exp H_{i,0} \cup \exp H_{i,1}$. Since $\rho_G^{-1}(w_i)$ is connected and some element of it squares to an element in $\exp H_{i,1}$ we have $K(w_i) \subset \exp H_{i,1}$.

To see that the squaring map

$$S_q : \rho_G^{-1}(w_i) \longrightarrow \exp H_{i,1}$$

is surjective, let $Q_i = \{t \in T \mid w_i(t) = t^{-1}\}$. Then Q_i is a 1-dimensional subgroup of T which acts by left translations on $\rho_G^{-1}(w_i)$. The orbits are just the fibers of the squaring map. If G has rank r , $\dim T = r$ and $\exp H_{i,1}$ is homeomorphic to an $r-1$ torus. So rank $S_q = r-1$ and S_q is surjective. Q.E.D.

3. THE SPLITTING MACHINE

We shall develop a criterion for when N splits, i.e., when there is a homomorphic cross section α

$$0 \longrightarrow T \longrightarrow N \underset{\alpha}{\overset{\rho}{\rightleftarrows}} W \longrightarrow 1 .$$

If $\theta_1,\ldots,\theta_r$ is a set of simple roots for G with respect to T , then w is generated by reflections $w_1,\ldots,w_r$ and the relations are generated by

i) $w_i^2 = 1$

ii) $(w_i w_j)^{n_{ij}} = 1 \qquad n_{ij} \in \{2,3,4,6\}$.

So we look for $z_i \in \rho^{-1}(w_i)$ satisfying these same two relations. In the proof of Lemma 2 we introduced maps $\beta_i \colon S^3 \longrightarrow G$. Let q_i be any choice of element in $\beta_i(M_i) \subset \rho^{-1}(w_i)$. Notice that

i)′ $q_i^2 = \exp\left(\frac{q_i}{2}\right)$.

Lemma 3: (J. Tits)

ii)′ $\underbrace{q_i q_j \cdots}_{n_{ij} \text{ terms}} = \underbrace{q_j q_i \cdots}_{n_{ij} \text{ terms}}$.

Proof: See [8] Proposition 3. The argument given there is for complex semi-simple Lie groups, but it can be easily translated into a result for real, compact, semi-simple Lie groups. Q.E.D.

Since W acts on $\mathfrak{L}_T$, the integral group ring $\mathbb{Z}w$ also acts on $\mathfrak{L}_T$.

Definition: $v_{ij} \in \mathbb{Z}w$ is equal to

$$1 - w_j + w_i w_j + \ldots (-1)^{n_{ij}-1} \underbrace{\ldots w_i w_j}_{n_{ij} \text{ factors}}$$

Theorem 3: N splits ⇔ there exists elements $X_1, \ldots, X_r$ in $\mathfrak{L}_T$ such that

a) $2X_i = (\theta_i(X_i) - \tfrac{1}{2})\tau_i \mod I$

b) $v_{ij}(X_i) \equiv v_{ji}(X_j) \mod I$.

Proof: Assume $X_1, \ldots, X_r$ exist. Let $z_i = \exp X_i$, and let $\alpha(w_i) = z_i q_i$. Then $(\alpha(w_i))^2 = z_i q_i z_i q_i = z_i(q_i z_i q_i^{-1}) q_i^2 =$

$z_i w_i(z_i) \exp\left(\frac{\tau_i}{2}\right) = \exp\left(X_i + w_i(X_i) + \frac{\tau_i}{2}\right) = \exp\left(2X_i - \theta_i(X_i)\tau_i + \frac{\tau_i}{2}\right) = 0$ by (a) . A similar straight forward computation shows that (a) and (b) imply $(\alpha(w_i)\alpha(w_j))^{n_{ij}} = 1$.

Conversely, if α is a section we let X_i be any element in $\exp^{-1}(\alpha(w_i)q_i)$. It is easy to show that since the $\alpha(w_i)$ satisfy i) and ii) the X_i will satisfy a) and b). Q.E.D.

Remark: In [9] more conceptual conditions will be given which determine when $0 \longrightarrow T \longrightarrow N \longrightarrow W \longrightarrow 1$ splits.

4. SAMPLE COMPUTATIONS

We give three sample applications of Corollary 4. All of the results in Theorem 2 can be gotten by similar computations. (Also see [9]).

Let $\overline{\theta}_1, \overline{\theta}_2, \ldots, \overline{\theta}_r \in \mathfrak{L}_T$ be a dual basis to $\theta_1, \theta_2, \ldots, \theta_r$. If G has trivial center, then $\overline{\theta}_1, \overline{\theta}_2, \ldots, \overline{\theta}_r$ is a basis for I . If $X = \Sigma x_i \overline{\theta}_i$ we shall write $X = (x_1, x_2, \ldots, x_r)$. Let $c_{ij} = \theta_i(\tau_j)$ which is an integer. The c_{ij} are called the Cartan integers for G . Notice that $\tau_j = (c_{1j}, c_{2j}, \ldots, c_{rj})$. Let $X_i = (x_{i1}, \ldots, x_{ir})$. We shall need the following formulas:

1) If $n_{ij} = 2$, then $v_{ij}(X_i) = x_{ij}\tau_j$

2) If $n_{ij} = 3$, then $v_{ij}(X_i) = X_i - (x_{ii} + x_{ij})\tau_i$

3) If $n_{ij} = 4$ and $c_{ij} = -1$, then $v_{ij}(X_i) = 2(x_{ii} + x_{ij})\tau_j$

4) If $n_{ij} = 4$ and $c_{ij} = -2$, then $v_{ij}(X_i) = (x_{ii} + 2x_{ij})\tau_j$.

Application I: If G is the unique group with local type F_4 , then $T_G \longrightarrow N_G \longrightarrow W_G$ does not split.

The Dynkin diagram for F_4 is

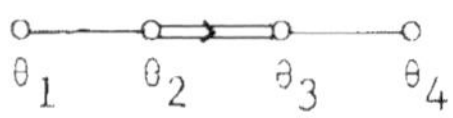

and its Cartan matrix (c_{ij}) is

$$\begin{pmatrix} 2 & -1 & 0 & 0 \\ -1 & 2 & -2 & 0 \\ 0 & -1 & 2 & -1 \\ 0 & 0 & -1 & 2 \end{pmatrix} .$$

If $T_G \longrightarrow N_G \longrightarrow W_G$ splits, then there exist elements $X_1, X_2, X_3, X_4 \in \mathfrak{L}_{T_{F_4}}$ which satisfy (a) and (b) of Corollary 4. Condition (a) for X_2 implies that

$$2(x_{21}, x_{22}, x_{23}, x_{24}) \equiv (x_{22}-\tfrac{1}{2})(-1,2,-1,0) \bmod I .$$

In particular then

$$2x_{23} \equiv (x_{22}-\tfrac{1}{2})(-1) \bmod 1 \quad \text{or} \quad (*)\, 2x_{23} + x_{22} \equiv +\tfrac{1}{2} \bmod 1 .$$

Next consider condition (b) for $i = 2,\ j = 3$. We have $n_{ij} = 4$ and formulas (3) and (4) apply. This gives

$$(x_{22} + 2x_{23})(0,-2,2,-1) \equiv 2(x_{33} + x_{32})(-1,2,-1,0) .$$

So $(x_{22} + 2x_{23})(-1) \equiv 0 \bmod 1$, contradicting $(*)$.

Note that the same proof applies to $Sp(n)/\text{center}$, $(n \geq 3)$, since the first vertex of the Dynkin diagram is not involved in the argument. The normalizers in $Sp(1)/\text{center}$ and $Sp(2)/\text{center}$ split, being isomorphic respectively to $SO(3)$ and $SO(5)$ (see next section).

<u>Application II</u>: $T_{SO(\text{odd})} \longrightarrow N_{SO(\text{odd})} \longrightarrow W_{SO(\text{odd})}$ splits.

We shall just do the computations for $SO(5)$. It will be obvious how to generalize.

The Cartan matrix is $\begin{pmatrix} 2 & -2 \\ -1 & 2 \end{pmatrix}$ and $n_{ij} = 4$, so formulas (3) and (4) again apply.

We have three conditions on X_1, X_2 .

A. $2(x_{11}, x_{12}) \equiv (x_{11} - \frac{1}{2})(2,-1) \bmod I$

B. $2(x_{21}, x_{22}) \equiv (x_{22} - \frac{1}{2})(-2,2) \bmod I$

C. $(x_{11} + 2x_{12})(-2,2) \equiv 2(x_{22} + x_{21})(2,-1) \bmod I$.

These conditions are satisfied by $X_1 = (0, \frac{1}{4})$ and $X_2 = (\frac{1}{2}, 0)$.

<u>Application III</u>: If $ad(E_6)$ is the unique group with trivial center and local type E_6 , then $T_{ad(E_6)} \longrightarrow N_{ad(E_6)} \longrightarrow W_{ad(E_6)}$ does not split.

The Dynkin diagram for E_6 is

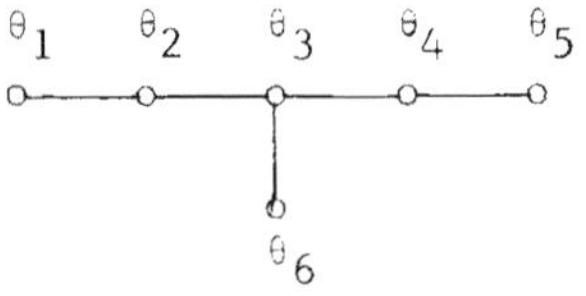

The Cartan matrix is

$$\begin{pmatrix} 2 & 0 & -1 & 0 & 0 & 0 \\ 0 & 2 & 0 & -1 & 0 & 0 \\ -1 & 0 & 2 & -1 & 0 & 0 \\ 0 & -1 & -1 & 2 & -1 & 0 \\ 0 & 0 & 0 & -1 & 2 & -1 \\ 0 & 0 & 0 & 0 & -1 & 2 \end{pmatrix}$$

If $T_{ad(E_6)} \longrightarrow N_{ad(E_6)} \longrightarrow W_{ad(E_6)}$ splits, then there exist elements $X_1, X_2, \ldots, X_6$ which satisfy (a) and (b) of Corollary 4. (a) implies

$$(a)' \quad X_i = \left(\frac{x_{ii}}{2} - \frac{1}{4}\right)\tau_i + \varepsilon_i \quad \text{where} \quad \varepsilon_i \in \tfrac{1}{2}I \quad \text{and} \quad \varepsilon_{ii} = \tfrac{1}{2} .$$

Furthermore formulas (1) and (2) can be rewritten as

(1)′ If $n_{ij} = 2$, then $v_{ij}(X_i) = \varepsilon_{ij}\tau_j$ (when $n_{ij} = 2$, then $\tau_{ij} = 0$) .

(2)′ If $n_{ij} = 3$, then $v_{ij}(X_i) = \varepsilon_i - (\frac{1}{2}+\varepsilon_{ij})\tau_i$ (when $n_{ij} = 3$, then $\tau_{ij} = -1$) .

(b) plus (1)′ applied to the pair X_2, X_3 yields

$$\varepsilon_{23}(-1,0,2,-1,0,0) \equiv \varepsilon_{32}(0,2,0,-1,0,0) \bmod I .$$

In particular $\varepsilon_{23} = 0$. Similarly, (b) plus (1)′ applied to the pairs X_2, X_5 and X_3, X_5 yield $\varepsilon_{25} = \varepsilon_{35} = 0$. (b) plus (2)′ applied to the pair X_4, X_2 yields that

$$\tau_4 - (\tfrac{1}{2}+\varepsilon_{42})\tau_4 \equiv \varepsilon_2 - (\tfrac{1}{2}+\varepsilon_{24})\tau_2 \bmod I .$$

The fifth component of this equation plus the fact that $\varepsilon_{25} = 0$ implies that

$$\text{(i)} \quad \varepsilon_{45} \not\equiv \varepsilon_{42} \bmod 1 .$$

The third component plus the fact that $\varepsilon_{23} = 0$ implies that

$$\text{(ii)} \quad \varepsilon_{43} \not\equiv \varepsilon_{42} \bmod 1 .$$

Similarly, the fifth component of the equation that we get by applying (b) and (2)′ to X_4, X_3 implies

$$\text{(iii)} \quad \varepsilon_{45} \not\equiv \varepsilon_{43} \bmod 1 .$$

Since ε_{45}, ε_{43}, and ε_{42} are of order 2, (i), (ii), and (iii) can not all be true. Therefore $T_{ad(E_6)} \longrightarrow N_{ad(E_6)} \longrightarrow W_{ad(E_6)}$ does not split. The same argument applies verbatim to E_7/center and E_8 .

APPENDIX 1

We include here a proof that the transformation groups (T,W) are the same for $Sp(n)$ and $SO(2n+1)$, since we have not found a convenient reference for this fact.

$Sp(n)$ is the group of $n \times n$ quaternionic matrices M such that $M\overline{M}^T = I$. A maximal torus for $Sp(n)$ is the subgroup of matrices

$$\Delta(\theta_1, \ldots, \theta_n) = \begin{bmatrix} e^{i\theta_1} & & & & & \\ & \cdot & & & 0 & \\ & & \cdot & & & \\ & & & \cdot & & \\ & 0 & & & \cdot & \\ & & & & & e^{i\theta_n} \end{bmatrix} .$$

The Weyl group is generated by:

$$\sigma_\ell : \Delta(\theta_1, \ldots, \theta_n) \longmapsto \Delta(\theta_1, \ldots, \theta_{\ell-1}, \theta_{\ell+1}, \theta_\ell, \theta_{\ell+2}, \ldots, \theta_n)$$

for $1 \leq \ell \leq n-1$ and

$$\tau : \Delta(\theta_1, \ldots, \theta_n) \longmapsto \Delta(\theta_1, \ldots, \theta_{n-1}, -\theta_n) .$$

σ_ℓ is induced by conjugation by

$$\begin{bmatrix} 1 & & & & & & \\ & \ddots & & & & & \\ & & 1 & & & & \\ & & & 0 & 1 & & \\ & & & 1 & 0 & & \\ & & & & & 1 & \\ & & & & & & \ddots \\ & & & & & & & 1 \end{bmatrix} \begin{matrix} \\ \\ \\ \text{row } \ell \\ \text{row } \ell+1 \\ \\ \\ \\ \end{matrix} ,$$

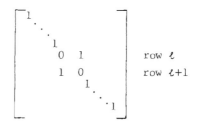

and τ is induced by conjugation by

$$\begin{bmatrix} 1 & & & & \\ & \cdot & & 0 & \\ & & \cdot & & \\ & & & \cdot & \\ & & & & 1 \\ & 0 & & & & j \end{bmatrix}.$$

$SO(2n+1)$ is the group of real $(2n+1) \times (2n+1)$ matrices M such that $MM^T = I$ and has a maximal torus consisting of the matrices

$$\Delta'(\theta_1,\ldots,\theta_n) = \begin{bmatrix} \cos\theta_1 & -\sin\theta_1 & & & & \\ \sin\theta_1 & \cos\theta_1 & & & & \\ & & \ddots & & & \\ & & & \cos\theta_n & -\sin\theta_n & \\ & & & \sin\theta_n & \cos\theta_n & \\ & & & & & 1 \end{bmatrix}.$$

The Weyl group for $SO(2n+1)$ is generated by

$$\sigma'_\ell : \Delta'(\theta_1,\ldots,\theta_n) \longmapsto \Delta'(\theta_1,\ldots,\theta_{\ell-1},\theta_{\ell+1},\theta_\ell,\theta_{\ell+2},\ldots,\theta_n)$$

for $1 \leq \ell \leq n-1$ and

$$\tau' : \Delta'(\theta_1,\ldots,\theta_n) \longmapsto \Delta'(\theta_1,\ldots,\theta_{n-1},-\theta_n) .$$

σ'_ℓ is induced by conjugation by

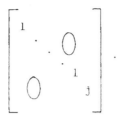

$$\begin{bmatrix} 1 & & & & & & & & & \\ & \ddots & & & & & & & & \\ & & 1 & & & & & & & \\ & & & 0 & 0 & 1 & 0 & & & \\ & & & 0 & 0 & 0 & 1 & & & \\ & & & 1 & 0 & 0 & 0 & & & \\ & & & 0 & 1 & 0 & 0 & & & \\ & & & & & & & 1 & & \\ & & & & & & & & \ddots & \\ & & & & & & & & & 1 \end{bmatrix} \begin{matrix} \\ \\ \\ \text{row } 2\ell-1 \\ \\ \\ \text{row } 2\ell+2 \\ \\ \\ \\ \end{matrix}$$

and τ' is induced by conjugation by

$$\begin{bmatrix} 1 & & & & & \\ & \ddots & & & & \\ & & 1 & & & \\ & & & 0 & 1 & \\ & & & 1 & 0 & \\ & & & & & -1 \end{bmatrix} .$$

Clearly the isomorphism of tori

$$\Delta(\theta_1, \ldots, \theta_n) \longrightarrow \Delta'(\theta_1, \ldots, \theta_n)$$

is equivariant with respect to the isomorphism of Weyl groups generated by

$$\sigma_\ell \longmapsto \sigma_\ell' , \quad \tau \longmapsto \tau' .$$

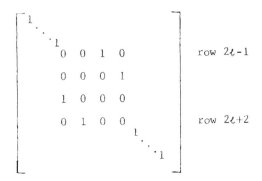

APPENDIX 2

Cohomology of Lie Groups

If one views the Weyl group W of a Lie group W as a degenerate form of G, then it is natural to attempt to compute the topological invariants of G in terms of W. For example, we have the beautiful result of Borel-Leray that $H^*(BG,\mathbb{Q}) \cong H^*(BT,\mathbb{Q})^W$. (Rector has observed that if p does not divide the order of the Weyl group; then the analogous result is true with $\mathbf{Z}/p\mathbf{Z}$ coefficients.)

Unfortunately W does not completely determine even the local isomorphism of G. Theorem 1 suggests that we attempt to compute the topological invariants of G in terms of N.

Proposition A.1: For any ring R, $H^*(BG,R)$ is a direct summand of $H^*(BN,R)$.

Proof: Given any differentiable fibre bundle $F \longrightarrow E \xrightarrow{\pi} B$ Becker and Gottlieb [2] have defined a trace homomorphism $\pi^+ : H^*(E) \longrightarrow H^*(B)$ such that $\pi^+\pi^* = \chi(F)\cdot \mathrm{Id}_{H^*(B)}$ where $\chi(F)$ is the Euler characteristic of F. (Borel [3] defined π^+ under the assumption that F is orientable.) $\chi(G/N) \times$ order of $W = \chi(G/T) =$ order of W. Thus $\chi(G/N) = 1$ and the proposition is proven by applying the Becker-Gottlieb result to the fibration $G/N \longrightarrow BN \longrightarrow BG$. Q. E. D.

The following results show that W almost determines G.

Proposition A.2: [see [9]]

(a) If G is simple and p is odd, then $BG_{(p)}$ (localization in the sense of Sullivan) is determined by W as a transformation group of T.

(b) $H^*(BN,\mathbb{Z}[\frac{1}{2}]) \simeq H^*(B(T\times_W W), \mathbf{Z}[\frac{1}{2}])$

(c) Assume $\pi_1 G_1 = \pi_1 G_2 = (1)$. Then $G_1 = G_2 \Leftrightarrow W_{G_1} = W_{G_2}$ as transformation groups of maximal tori.

Remarks:

1. (a) is probably true for all connected, compact Lie groups.
2. (b) is a consequence of Lemma 3.
3. (c) is probably well known. A proof of (c) is implicit in the proof of [5] Chapter VI #2 Proposition 9.

The following questions are suggested by (b).

1. Does the Serre spectral sequence for $BT \longrightarrow B(T\times_{\alpha} W) \longrightarrow K(W,1)$ collapse?
2. More generally, given a fibration $F \longrightarrow E \longrightarrow B$ with cross section does the Serre spectral sequence collapse even when the action of $\pi_1 B$ on $H^*(F)$ is nontrivial?
3. Does the Serre-Hochschild spectral sequence collapse when it is applied to a semi-direct product of finite groups?

REFERENCES

[1] F. Adams; Lectures on Lie groups, W. A. Benjamin, 1969.

[2] J. Becker. D. Gottlieb; Coverings of fibrations, Compositio Mathematica. vol. 26(2), (1973) 119-128.

[3] A. Borel, Sur la torsion de groupes de Lie, J. Math. Pures, Appl. (9) 35(1956) 127-139.

[4] R. Bott, H. Samelson, Applications of the theory of Morse to symmetric spaces, Amer. J. of Math. LXXX(1958) 964-1029.

[5] N. Bourbabi, Groupes et algèbres de Lie, chapitres 4,5 et 6, Hermann 1968.

[6] S. Helgason; Differential Geometry and Symmetric Spaces, Academic Press 1962.

[7] H. Hopf, Maximale Toroide und singulare Elemente in geschlossen Lieschen Gruppe. Comment. Math. Helv. 15(1943) 59-70.

[8] J. Tits, Sur les constantes de structure et le théorème d'existence de algèbres de Lie semi-simples. Inst. Hautes Etudes Sci. Publ. Math. No. 31 (1966) 21-58.

[9] B. Williams. Weyl groups and the cohomology of Lie groups. to appear.

Rice University

METASTABLE EMBEDDING AND 2-LOCALIZATION

Henry H. Glover and Guido Mislin*

Ohio State University

ETH Zurich

Introduction

Throughout this paper let M^n denote a closed smooth manifold of dimension n. Let M_0 denote the complement of a point in M. We prove the following results.

Theorem 0.1. If $\tilde{H}_i(M;\mathbb{Z}/2) = 0$ for $i \leq k$, then

A. M^n smoothly embeds in $R^{\max([3n/2]+2,\, 2n-k)}$, if k is even.

B. If, in addition, M_0 immerses in R^{2n-2j} for some $j \leq k$, then M smoothly embeds in $R^{\max([3n/2]+2,\, 2n-2j+1)}$.

Theorem 0.1.A and the following corollary were first proved in [9] by different methods, c.f. [10] and [2].

We call M^n a $\mathbb{Z}/2$-homology sphere if $H_*(M^n;\mathbb{Z}/2) \cong H_*(S^n;\mathbb{Z}/2)$.

Corollary 0.2. Every $\mathbb{Z}/2$-homology sphere smoothly embeds in $R^{[3n/2]+2}$.

Notice that this corollary implies the metastable embedding of odd spherical space forms.

Theorem 0.1.B is a generalization of [1]. It has the following corollaries.

We call manifolds M and N $\bar{2}$-equivalent if their $\bar{2}$-localizations are homotopy equivalent $M_{\bar{2}} \simeq N_{\bar{2}}$ (see [2] and [6]). Note that this functor localizes the higher homotopy groups and is a special case of a functor in [4].

* Talk delivered by Henry Glover.

Corollary 0.3. Let M^n and N^n be $\bar{2}$-equivalent manifolds. Suppose that 1) there is a k such that $\tilde{H}_i(M;\mathbb{Z}/2) = 0$ for $i \leq k$, 2) there is a $j \leq k$ such that M immerses in R^{2n-2j}. Then N smoothly embeds in $\mathbb{R}^{\max([3n/2]+2,\, 2n-2j+1)}$.

Corollary 0.4. Let M^n be $\bar{2}$-equivalent to a π-manifold. Suppose that $\tilde{H}_i(M;\mathbb{Z}/2) = 0$ for $i \leq k$. Then M^n smoothly embeds in $R^{\max([3n/2]+2,\, 2n-2k+1)}$.

We here note that R. Rigdon shows in [10] that manifolds $\bar{2}$-equivalent by virtue of a global map embed in the same metastable dimension.

The plan of the paper follows.

In section 1 we recall a main result of [8] and [4] about localization and the main result of [6] and [2] about metastable immersion.

In section 2 we prove the needed results about the co-connectivity of the symmetric deleted product of M and an associated pair. The proof uses the Smith sequences.

In section 3 we prove the main results.

1. Localization and immersion

We will use the notation of [8] and [6]. If X is a connected nilpotent CW homotopy type then X_p denotes its p-localization (p a prime or 0); there are canonical maps $X \longrightarrow X_p$ respectively $X_p \longrightarrow X_0$. We will need the following basic result of [8] and [4].

Proposition 1.1. Let W be a connected finite CW complex and let X be a connected nilpotent CW complex of finite type. Then the set of pointed homotopy classes $[W, X]$ is the pullback of the diagram of sets $\{[W, X_p] \longrightarrow [W, X_0] \mid p \in P\}$, where P denotes the set of primes.

We will use from [6] and [2] the following result.

Proposition 1.2. Let M^n and N^n be manifolds of dimension n whose $\bar{2}$-localizations $M_{\bar{2}}$ and $N_{\bar{2}}$ are homotopy equivalent. Suppose M^n immerses in R^{n+k} for some $k \geq [n/2]+1$. Then N^n immerses in $R^{n+2[k/2]+1}$.

Corollary 1.3. Let M^n and N^n be as in 1.2 and assume that M is a π-manifold. Then N immerses in $R^{n+2[(n+2)/4]+1}$.

2. Smith theory and the deleted product

Let $M^* = M \times M - \Delta$ denote the deleted product of M with itself. It is equipped with an involution $\iota : M^* \longrightarrow M^*$ given by interchanging factors. We denote by M^*/ι the orbit space of this action.

Proposition 2.1. If $\tilde{H}_i(M;\mathbb{Z}/2) \cong 0$ for $i \leq k$ then $H^j(M^*;R) \cong 0$ for $j \geq 2n-k$ and for $R = Q$, $\mathbb{Z}/2$ or $\mathbb{Z}_2$ (the integers localized at 2).

Proof. Let $M_0 = M - \{\text{point}\}$. One has a fibration $M_0 \longrightarrow M^* \longrightarrow M$ induced by a projection. Clearly $\tilde{H}_i(M;\mathbb{Z}/2) = 0$ for $i \leq k$ implies that $H^{n-i}(M_0;\mathbb{Z}/2) = 0$ for $i \leq k$. Therefore $H^j(M^*;\mathbb{Z}/2) \cong 0$ for $j \geq 2n-k$ from the fibration above. Clearly this implies that $H^j(M^*;R) \cong 0$ for $j \geq 2n-k$ and $R = Q$ or $\mathbb{Z}_2$. Hence the result.

Corollary 2.2. If $\tilde{H}_i(M;\mathbb{Z}/2) = 0$ for $i \leq k$, then $H^j(M^*/\iota;R) = 0$ for $j \geq 2n-k$ and for $R = Q$, $\mathbb{Z}/2$ or $\mathbb{Z}_2$.

Proof. Again it is obviously sufficient to prove the result for $R = \mathbb{Z}/2$. Observe that M^* has the equivariant homotopy type of a finite complex (with cellular action). This can be seen by removing a suitable open neighborhood of the diagonal in $M \times M$, instead of removing Δ. Therefore we have the Smith sequence [3].

$$\begin{array}{ccc} & H^*(M^* \; ; \; \mathbb{Z}/2) & \\ \nearrow & & \searrow \\ H^*(M^*/\iota \; ; \; \mathbb{Z}/2) & \xleftarrow{\;\delta\;} & H^*(M^*/\iota \; ; \; \mathbb{Z}/2) \end{array}$$

with δ a map of degree $+1$. Applying the previous proposition we see that $H^j(M^* \; ; \; \mathbb{Z}/2) = 0$ for $j \geq 2n - k$. Since M^*/ι has the homotopy type of a finite CW-complex, $H^j(M^*/\iota \; ; \mathbb{Z}/2) = 0$ for j big enough. Using a decreasing induction on j in the Smith sequence we obtain the desired result for $R = \mathbb{Z}/2$. Clearly knowing the result for $R = \mathbb{Z}/2$ implies the result for $R = Q$ and $\mathbb{Z}_2$.

The following proposition is a modification of [1, lemma 2.7]. Denote by $S_\epsilon(M_0) \subset M_0 \times M_0$ the normal sphere bundle of the diagonal M_0 of $M_0 \times M_0$; notice that $S_\epsilon(M_0)$ is equivalent to the tangent sphere bundle of M_0.

<u>Proposition 2.3.</u> If $\tilde{H}_i(M \, ; \mathbb{Z}/2) = 0$ for $i \leq k < n/2$, then $H^j(M_0^* \, , S_\epsilon(M_0) \; ; R) = 0$ for $j \geq 2n - 2k - 1$ and for $R = Q$, $\mathbb{Z}/2$ or $\mathbb{Z}_2$.

<u>Proof.</u> Again we can restrict to the case $R = \mathbb{Z}/2$. By excision and the exact cohomology sequence of the pair $(M_0 \times M_0 \, , \Delta M_0)$

$$H^j(M_0^* \, , S_\epsilon(M_0) \; ; \mathbb{Z}/2) \cong H^j(M_0 \times M_0 \; ; \mathbb{Z}/2) \quad \text{if } j \geq n \, .$$

Notice that $M_0 \times M_0 = M \times M - M \vee M$. Hence

$$H^j(M_0 \times M_0 \; ; \mathbb{Z}/2) \cong H^j(M \times M - M \vee M \; ; \mathbb{Z}/2)$$

$$\cong H_{2n-j}(M \times M \, , M \vee M \, ; \mathbb{Z}/2)$$ by Lefschetz duality. Now

$$H_{2n-j}(M \times M \, , M \vee M \, ; \mathbb{Z}/2) \cong \tilde{H}_{2n-j}(M \wedge M \, ; \mathbb{Z}/2) = 0 \quad \text{for } j \geq 2n - 2k - 1,$$

since $2n - j \leq 2n - (2n - 2k - 1) = 2k + 1$ and $\tilde{H}_i(M \, ; \mathbb{Z}/2) = 0$ for $i \leq k$. Hence the result.

By passing to the orbit spaces for the obvious $\mathbb{Z}/2$ actions, we get as in corollary 2.2.

<u>Corollary 2.4.</u> If $k \geq 1$ and $\tilde{H}_i(M ; \mathbb{Z}/2) = 0$ for $i \leq k < n/2$, then $H^j(M_0^*/\iota : S_\epsilon(M_0)/\iota ; R) = 0$ for $j \geq 2n - 2k - 1$ and for $R = Q$, $\mathbb{Z}/2$ or $\mathbb{Z}_2$.

We recall the following result from [1]. Let X and Y be spaces with involutions. Denote by $E(X, Y)$ the set of equivariant maps from X to Y.

<u>Proposition 2.5.</u> Let Y be a topological space equipped with an involution ι_Y. If $E(M_0^*, Y) \neq \emptyset$ then $E(M^*, \Sigma Y) \neq \emptyset$.

<u>Proof.</u> Notice that $M^* = M_0^* \cup (M_0 \amalg M_0)$. Let $\varphi \in E(M_0^*, Y)$. Since the canonical inclusion can: $Y \dashrightarrow \Sigma Y$ is null homotopic (as a map) there exists an extension of $\mathrm{can} \circ \varphi : M_0^* \longrightarrow \Sigma Y$ (as a map) to a map $\varphi_1 : M_0^* \cup M_0 \dashrightarrow \Sigma Y$. Finally define $\tilde{\varphi} \in E(M^*, \Sigma Y)$ by

$$\tilde{\varphi}(x, y) = \begin{cases} \varphi_1(x_1, x_2) & , \text{ if } (x_1, x_2) \in M_0^* \cup M_0 \\ \iota_y \circ \varphi_1(x_2, x_1) , & \text{otherwise.} \end{cases}$$

We check that $\tilde{\varphi}$ is continuous and equivariant. Hence the result.

3. <u>The proof of the main theorems</u>

Let W be a connected finite CW-complex and let X and Y be connected nilpotent complexes of finite type.

<u>Proposition 3.1.</u> The lifting problem

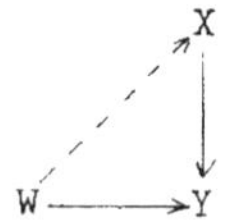

has a solution provided (i) the lifting problem

$$\begin{array}{ccc} & & X_p \\ & \nearrow & \downarrow \\ W & \longrightarrow & Y_p \end{array}$$

has a solution for all primes p and (ii) the lifting problem

$$\begin{array}{ccc} & & X_0 \\ & \nearrow & \downarrow \\ W & \longrightarrow & Y_0 \end{array}$$

has a solution compatible with those above. This means that

$$\begin{array}{ccccc} & & X_p & \longrightarrow & X_0 \\ & \nearrow & \downarrow & \nearrow & \downarrow \\ W & \longrightarrow & Y_p & \longrightarrow & Y_0 \end{array}$$

commutes for all $p \in P$.

The proof of proposition 3.1 follows immediately from proposition 1.1.

<u>The proof of theorem 0.1.A.</u> Note that by Haefliger's theorem [7] it suffices to solve the following lifting problem

$$(\mathrm{I}) \qquad \begin{array}{ccc} & & \mathbb{R}P^{m-1} \\ & \nearrow & \downarrow \text{can} \\ M^*/\iota & \xrightarrow[u]{} & \mathbb{R}P^{\infty} \end{array}$$

where $m = \max([3n/2] + 2 \ , \ 2n - k)$. Here $u : M^*/\iota \longrightarrow \mathbb{R}P^{\infty}$ is the classifying map for the double covering $M^* \longrightarrow M^*/\iota$. To solve (I) we consider the following lifting problem

$$(\mathrm{II}) \qquad \begin{array}{ccc} & & \mathbb{R}P^{2n-k-1} \\ & \overset{\phi}{\nearrow} & \downarrow \\ M^*/\iota & \xrightarrow[u]{} & \mathbb{R}P^{\infty} \end{array}$$

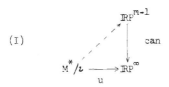

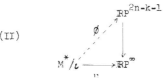

Clearly if (II) can be solved then so can (I). We have assumed k even. We can then apply proposition 3.1 to (II) since the spaces fulfill the requirements of proposition 3.1 ; in particular $\mathbb{R}P^{2n-k-1}$ is simple since $2n - k - 1$ is odd. If p is an odd prime then $\mathbb{R}P^{\infty}_{p}$ is contractible and hence we can solve

$$\begin{array}{ccc} & & \mathbb{R}P^{2n-k-1}_{p} \\ & \overset{\phi(p)}{\nearrow} & \downarrow \\ M^{*}/\iota & \longrightarrow & \mathbb{R}P^{\infty}_{p} \end{array}$$

by choosing $\phi(p) = 0$. For the prime 2 we observe that the obstructions for the lifting

$$\begin{array}{ccc} & & \mathbb{R}P^{2n-k-1}_{2} \\ & \overset{\phi(2)}{\nearrow} & \downarrow \\ M^{*}/\iota & \longrightarrow & \mathbb{R}P^{\infty}_{2} \end{array}$$

lie in $H^{j}(M^{*}/\iota \ ; \pi_{j}(\mathbb{R}P^{\infty}_{2}, \mathbb{R}P^{2n-k-1}_{2}))$. Note $\pi_{j}(\mathbb{R}P^{\infty}_{2}, \mathbb{R}P^{2n-k-1}_{2}) = \pi_{j-1}\mathbb{R}P^{2n-2k-1}_{2}$, so $\pi_{j}(\mathbb{R}P^{\infty}_{2}, \mathbb{R}P^{2n-k-1}_{2}) = 0$ for $j \leq 2n-k-1$; for $j > 2n-k-1$ $\pi_{j}(\mathbb{R}P^{\infty}_{2}, \mathbb{R}P^{2n-k-1}_{2})$ is a 2-group or (in case $j = 2n-k$) $\mathbb{Z}_2$. Thus $H^{j}(M^{*}/\iota \ ; \pi_{j}(\mathbb{R}P^{\infty}_{2}, \mathbb{R}P^{2n-k-1}_{2})) = 0$ for all j by Corollary 2.2, giving the desired lift. Notice that $\mathbb{R}P^{2n-k-1}_{0} \simeq K(Q, 2n-k-1)$ since k is even, so that the canonical map $\rho : \mathbb{R}P^{2n-k-1}_{2} \longrightarrow \mathbb{R}P^{2n-k-1}_{0}$ factors

$$\begin{array}{ccccc} & & K(\mathbb{Z}_2, 2n-k-1) & & \\ & \overset{\theta}{\nearrow} & & \searrow & \\ \mathbb{R}P^{2n-k-1}_{2} & & \xrightarrow{\quad \rho \quad} & & \mathbb{R}P^{2n-k-1}_{0} \end{array} .$$

It is a standard result of obstruction theory that there exists a map $\psi(2) : M^{*}/\iota \longrightarrow \mathbb{R}P^{2n-k-1}_{2}$ which agrees with $\phi(2)$ on the $2n-k-2$-skeleton of M^{*}/ι, and such that $\theta \circ \psi(2) \simeq 0$. As a result

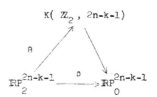

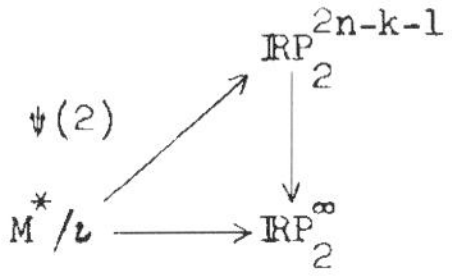

commutes, and $\rho \circ \psi(2) \simeq 0$. Now apply proposition 3.1 to complete the proof of theorem 0.1.A.

We here remark that theorem 0.1.A is also true for k odd (see [9], [2], [10]). However our proof then fails because $\mathbb{RP}^{2n-k-1}$ is then not a nilpotent space.

We conjecture that some modification of our technique will allow the application of proposition 3.1 to prove theorem 0.1.A for k odd.

We also conjecture that a proof using proposition 3.1 can be given for Rigdon's theorem [10] which states that manifolds $\bar{2}$-equivalent by means of a global map embed in the same metastable dimension.

<u>The proof of 0.1.B.</u> Again it is enough by Haefliger [7] to solve the lifting problem

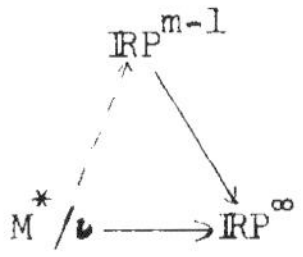

where $m = \max(\lceil 3n/2 \rceil + 2\ ,\ 2n - 2j + 1)$. Because M_0 immerses in $\mathbb{R}^{2n-2j}$, there exists a lift

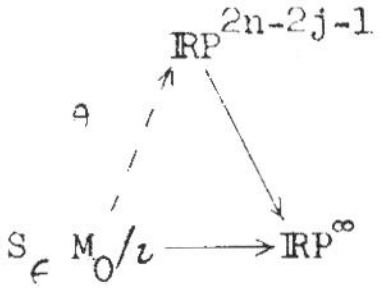

This lift gives a commutative diagram

$$\begin{array}{ccc} S_\epsilon M_0/\iota & \xrightarrow{\theta(2)} & \mathbb{R}P_2^{2n-2j-1} \\ \downarrow & \overset{\bar\theta(2)}{\nearrow} & \downarrow \\ M_0^*/\iota & \longrightarrow & \mathbb{R}P_2^\infty \end{array}$$

with the existence of $\bar\theta(2)$ still to be shown. That $\bar\theta(2)$ exists follows using that $\tilde{H}_i(M;\mathbb{Z}/2) = 0$ for $i \le k$. Without loss of generality assume that $k \le n/2$. This implies that all the groups

$$H^q(M_0^*/\iota, S_\epsilon M_0/\iota ; \pi_q(\mathbb{R}P_2^\infty, \mathbb{R}P_2^{2n-2j-1})) = 0 .$$

Namely for $q \le = 2n - 2j - 1$ the coefficient group $\pi_q(\mathbb{R}P_2^\infty, \mathbb{R}P_2^{2n-2j-1}) = 0$ and for $q \ge 2n - 2j$, the result follows by Corollary 2.4.

As in the proof of theorem 0.1.A we can choose $\bar\theta(2)$ such that $\rho \circ \bar\theta(2) \simeq 0$, $\rho : \mathbb{R}P_2^{2n-2j-1} \longrightarrow \mathbb{R}P_0^{2n-2j-1}$ the canonical map. To construct

$$\begin{array}{ccc} & \mathbb{R}P^{2n-2j-1} & \\ \overset{\bar\theta}{\nearrow} & & \searrow \\ M_0^*/\iota & \longrightarrow & \mathbb{R}P^\infty \end{array}$$

we can find trivial lifts for all the odd primes and then put the lifts together using proposition 3.1. This will give us the desired lift $\bar\theta$. We then use a result of section 2. For this note $\bar\theta$ induces an equivariant map $\varphi : M_0^* \longrightarrow S^{2n-2j-1}$ and by proposition 2.5 there exists an equivariant map $\varphi : M^* \longrightarrow S^{2n-2j}$. Hence M^n embeds in

$$\mathbb{R}^{\max(\lceil 3n/2 \rceil + 2\ ,\ 2n-2j+1)} .$$

<u>The proof of corollary 0.3.</u> By proposition 1.2 N^n and hence N_0^n immerses in $\mathbb{R}^{2n-2j}$. Now apply theorem 0.1 B to get the result.

We here conjecture that Theorem 1.B can be strengthened to a $\bar{2}$-local analogue of [5]. This would improve the dimension by 2 and give the converse.

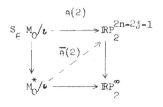

References

[1] J. C. Becker and H. H. Glover; Note on the embedding of manifolds in euclidean space, PAMS 27(1971) 405-410.

[2] M. Bendersky; Semi Localization, and a theorem of Glover and Mislin (appears in this proceedings).

[3] A. Borel; Seminar on Transformation Groups, Annals of Math. Studies no. 46.

[4] A. K. Bousfield and D. M. Kan; Homotopy Limits, Completions and Localizations, Springer Verlag Lecture Notes in Math. no. 304.

[5] Keith Ferland; Embeddings of k - orientable manifolds (to appear in the Michigan Mathematical Journal).

[6] Henry H. Glover and Guido Mislin; Immersion in the metastable range and 2-localization (to appear in PAMS).

[7] André Haefliger; Plongements différentiable dans le domain stable, Comm. Math. Helv. 37(1941) 155-176.

[8] P. Hilton, G. Mislin and J. Roitberg; Homotopical localization, Proc. London Math. Soc. (1973) 693-706.

[9] Elmer Rees; Embedding odd torsion manifolds, Bull. London Math. Soc. 3(1971) 356-362.

[10] R. Rigdon; p - Equivalences and embeddings of manifolds (to appear).

The mod 3 homotopy type of F_4

John R. Harper

In this paper we study the mod 3 homotopy type of the compact, simple, simply connected Lie group F_4. This is the simplest case of a Lie group with torsion whose mod p structure is unknown. We show

Theorem. There is a finite complex K such that F_4 is mod 3 equivalent to $K \times B_5(3)$.

Here $B_5(3)$ is the S^{11}-bundle over S^{15} classified by α_1. The mod 3 cohomology of K is

$$\Lambda(x_3, x_7) \otimes Z_3[x_8]/(x_8^3)$$

$$x_7 = \mathcal{P}^1 x_3, \quad x_8 = \beta x_7$$

Since K is a mod 3 retract of an H-space it follows that K is a mod 3 H-space, whose mod 3 cohomology is primitively generated for dimensional reasons. Applying the Zabrodsky mixing technique with ingredients F_4, Lie multiplication at primes $\neq 3$ and $K \times B_5(3)$, product multiplication (on the localization) at 3, we obtain a finite H-space with 3 torsion whose mod 3 cohomology is primitively generated. Now results of Browder [4] and Zabrodsky [8] assert that if a finite H-space is <u>homotopy associative</u> and its mod p cohomology (p odd) is primitively generated, then it is p-torsion free. Thus the Browder, Zabrodsky result cannot be further extended, cf problem 45 [9].

Our decomposition is related to another result of Browder [3]. He proves that if the mod p cohomology of a finite H-space is primi-

The author has been partially supported by NSF Grant GP38024.

tively generated, then the p-torsion is of order at most p. This too is best possible, even considering only simply connected spaces at odd primes.

The methods of Massey and Peterson [6] are extensively used here. Their results are for the prime 2. They can be extended in a straightforward manner to odd primes using a result of Barcus [1], c.f. [2] and [7].

The basic means of using this theory is as follows. Let X be a space whose mod p cohomology is of the form U(M). Form a projective resolution of M in the category of unstable modules over the Steenrod algebra. The key construction is a geometric realization of the projective resolution, [6] part III. This is a tower of fibre spaces E_i having several useful properties. One property is the existence of maps $X \to E_i$ which under favorable conditions are mod p cohomology isomorphisms through a range of dimensions which increases with i. This is useful because the Massey, Peterson, Barcus results enable one to compute the cohomology of E_i rather easily. Thus the tower can be used as a mod p Postnikov resolution of X.

We give two constructions of the complex K. The first uses the Massey-Peterson theory and indicates the uniqueness of the mod 3 homotopy type of K. The second is less computational, and developed at the conference through discussions with G. Mislin, J. Stasheff and A. Zabrodsky.

Let E_1 be the two-stage Postnikov system over $K(\mathbb{Z}, 3)$ with k-invariant $\mathcal{P}^3\mathcal{P}^1\iota_3$. An application of Barcus shows there are unique classes $e_1 \in H^{22}(E_1)$ and $e_2 \in H^{24}(E_1)$, mod 3 coefficients, which restrict to $\mathcal{P}^1\iota_{18}$ and $\beta\mathcal{P}^1\beta\iota_{18}$ respectively in the fibre, and e_2 is an integral class. Let E_2 be the fibre space over E_1 given by,

$$E_2 \longrightarrow E_1 \xrightarrow{e_1, e_2} K(Z_3,22)\times K(Z,24).$$

In dimensions ≤ 26, E_2 has the right mod 3 cohomology. One can take K to be the homology approximation to E_2 through dimensions 26. Then it is readily seen that the mod 3 cohomology is as desired.

For the second construction, let

$$f: S^7 \longrightarrow Sp(2)_{(3)}$$

be the compositeof a generator of $\Pi_7(Sp(2))$ followed by localization.

Prop. 1. There is an H-structure on S^7 such that f is an H-map.

pf. (Zabrodsky). A separation element argument shows that, allowing different H-structures on S^7, the obstruction to f being an H-map lies in

$$\Pi_{14}(Sp(2)_{(3)})/f_{\#}\Pi_{14}(S^7).$$

Now $\Pi_{14}(Sp(2)_{(3)}) = Z_3$, and $\alpha_2 \in \Pi_{14}(S^7)$ generates a cyclic subgroup of order 3. A look at the Postnikov system easily shows $f_{\#}(\alpha_2) \neq 0$, thus the displayed coset is 0.

To construct K, consider the map of projective planes induced by f, $$\hat{f}: P_2S^7 \longrightarrow P_2Sp(2)_{(3)}.$$

Embedding the second projective plane in the classifying space,

$$i : P_2\, Sp(2)_{(3)} \longrightarrow B\,Sp(2)_{(3)}$$

gives K as the $Sp(2)_{(3)}$ bundle over P_2S^7 induced by $i\,\hat{f}$.

We now turn to the decomposition of F_4. It is known that $H^*(F_4) = U(M)$ where M is generated by x_3 and x_{11} subject to relations $\beta x_3 = 0$, $\beta x_{11} = 0$, $\mathcal{P}^3\mathcal{P}^1 x_3 = 0$, $\beta\mathcal{P}^1 x_{11} = 0$. As algebras over the Steenrod algebra we have

$$H^*(F_4) \cong H^*(K) \otimes H^*(B_5(3)).$$

where here, as elsewhere, the implicit coefficients are mod 3. We tabulate the data for a minimal resolution of M below, separating the pieces which algebraically come from K or $B_5(3)$. Generators are listed as (s, t) where s is the homological degree and t the internal degree.

Enough information is given to determine E_2 of the Massey-Peterson spectral sequence in dimensions ≤ 27.

K	generator	defining relation
	(0, 3)	
	(1, 19)	$\mathcal{P}^3\mathcal{P}^1(0,3)$
	(2, 23)	$\mathcal{P}^1(1,19)$
	(2, 25)	$\beta\mathcal{P}^1\beta(1,19)$
	(3, 29)	$\beta\mathcal{P}^1\beta(2,23) - \mathcal{P}^1(2,25)$

(0,3) and (2,25) put in as integral classes.

$B_5(3)$	(0, 11)	
	(1, 16)	$\beta\mathcal{P}^1(0,11)$
	(1, 19)	$\mathcal{P}^2(0,11)$
	(1, 23)	$\mathcal{P}^3(0,11)$

$B_5(3)$ continued.

(1,27)	$\mathcal{P}^3\mathcal{P}^1(0,11)$
(2,20)	$\mathcal{P}^1(1,16) - \beta(1,19)$
(2,23)	$\mathcal{P}^1(1,19)$
(2,24)	$\mathcal{P}^2(1,16) - \beta(1,23)$
(2,28)	$\mathcal{P}^3(1,16) - \beta(1,27)$
(3,25)	$\mathcal{P}^1\beta(2,20) - \beta(2,24)$
(3,28)	$(\beta\mathcal{P}^1 + \mathcal{P}^1\beta)(2,23) + \mathcal{P}^2(2,20)$
(3,29)	$\mathcal{P}^1\beta(2,24) + \beta(2,28)$
(4,30)	$(\beta\mathcal{P}^1 + \mathcal{P}^1\beta)(3,25) + \beta(3,29)$

(o,11) and (1,16) are put in as integral classes.

In E_2 of the spectral sequence, (s, t) represents an element with filtration (row) s and stem (column) t - s. In a geometric realization the element (s, t) gives rise to a k-invariant at level s and dimension $t - s + 1$, $s > 0$.

The key technical point is a non-zero differential in the spectral sequence.

Prop. 2. $d_2(1, 27) = (3, 28)$.

Rather than prove this in the stated form we prove an equivalent fact about the cohomology of the spaces in a tower built from <u>some</u> of the data above. We take a geometric realization of the data (s, t) with $t - s \leq 23$,

$$F_4 \longrightarrow E_3 \longrightarrow E_2 \longrightarrow E_1 \longrightarrow K(Z,3) \times K(Z,11).$$

In particular we have the fibration

$$E_2 \longrightarrow E_1 \xrightarrow{\ \theta\ } \underbrace{(K_{22} \times \bar{K}_{24})}_{\text{from } k} \times \underbrace{(K_{19} \times K_{22} \times K_{23})}_{\text{from } B_5(3)}$$

where $K_n = K(Z_3, n)$ and $\bar{K}_n = K(Z, n)$ and the k-invariant $\theta^*(\iota_n)$ is determined by the datum (2, n+1) as in [6] prop 26.1. Furthermore $\mathcal{P}^3\mathcal{P}^1\iota_{11} \neq 0$ in $H^{27}(E_1)$ because we have not killed that class. The non-zero differential is equivalent to

Prop. 3. $$\theta^*((\beta\mathcal{P}^1 + \mathcal{P}^1\beta)\iota_{22} + \mathcal{P}^2\iota_{19}) = \mathcal{P}^3\mathcal{P}^1\iota_{11} \text{ in } H^{27}(E_1).$$

Here ι_{22} lies in the factor associated with $B_5(3)$. We shall prove this proposition after using it to obtain the decomposition of F_4.

In view of prop. 3, the Barcus theorem shows that except for a class in dimension 23 (corresponding to (3, 25)), the mod 3 cohomology of E_2 agrees with that of F_4 in dimensions ≤ 26. The class corresponding to (3, 25) is killed in the formation of E_3. Hence the obstructions to lifting a map into E_3 to one into F_4 are mod 3 in dimensions ≥ 27. Let $f_1: B_5(3)_{(3)} \to K(Z, 11)$ and $f_2: K_{(3)} \to K(Z,3)$ be the obvious maps. Then we have

$$f_1 \vee f_2 : B_5(3)_{(3)} \vee K_{(3)} \longrightarrow K(Z,11) \times K(Z,3)$$

and f_1^*, f_2^* are mod 3 epimorphisms. There are no obstructions to lifting F_1 to F_4, by dimensional arguments. The possible obstructions to lifting f_2 are associated with (2, 20), (2, 24) and (3, 25). However in each of these cases the indeterminacy absorbs the obstruction, e.g. for (2, 20)

$$H^{19}(K_{(3)}) = \beta H^{18}(K_{(3)})$$

since

$$x_3 x_8^2 = \beta(x_3 \cdot x_7 \cdot x_8)$$

Thus we obtain a map

$$B_5(3)_{(3)} \vee K_{(3)} \longrightarrow F_4$$

which extends to a mod 3 cohomology isomorphism

$$B_5(3)_{(3)} \times K_{(3)} \longrightarrow F_4$$

We now prove prop. 3. First some notation. Let $e_1 = \theta^*(\iota_{19})$ $e_2 = \theta^*(\iota_{22})$. These can be thought of as universal examples for secondary operations defined by relations (2, 20) and (2, 23) respectively. Prop. 3 asserts that

$$(\beta P^1 + P^1\beta)e_2 + P^1 e_1 = P^3 P^1 \iota_{11}$$

in $H^{27}(E_1)$. We show that this relation is a consequence of Liulevicius decomposition of P^3, the mod 3 Hopf invariant one result [5]. The k-invariants corresponding to (1, 16) and (1, 19) from $B_5(3)$ can be factored

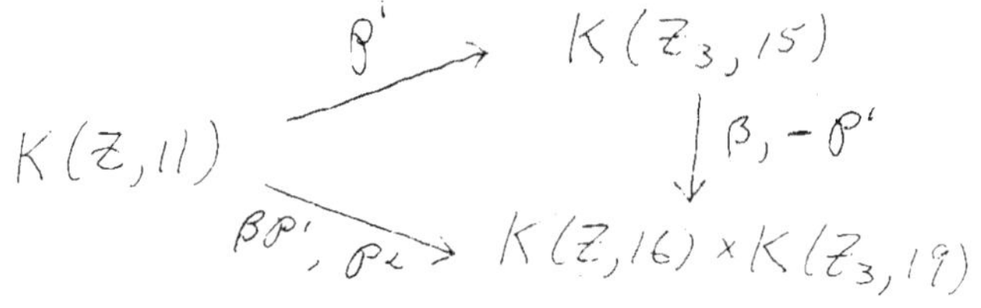

Let G be the two stage system over $K(Z_3, 15)$ with k-invariants $\beta\iota_{15}$ and $-P^1\iota_{15}$ as given by the vertical map. A map $g\colon E_1 \to G$ is induced which, because the k-invariants involved are loop classes, can be taken as an H-map. Let $r \in H^{23}(G)$, $\lambda \in H^{26}(G)$ be the classes defined by the relations

$$r\colon\ P^2\beta + (\beta P^1 + P^1\beta)(-P^1) = 0$$

$$\lambda\colon\ -P^2(-P^1) = 0 \ .$$

Using the fundamental sequence [6] and the fact that g is an H-map, we find,

$$g^*(r) = -\mathcal{P}^1 e_1 + \beta e_2$$

$$g^*(\lambda) = \mathcal{P}^1 e_2$$

Thus

$$g^*(\mathcal{P}^1 r + \beta\lambda) = \mathcal{P}^2 e_1 + (\beta\mathcal{P}^1 + \mathcal{P}^1\beta) e_2$$

But Liulevicius theorem asserts, [5] p.84,

$$\mathcal{P}^1 r + \beta\lambda = \mathcal{P}^3 c_{15}$$

with 0 indeterminacy in $H^{27}(G)$. Since

$$g^* \mathcal{P}^3 c_{15} = \mathcal{P}^3 \mathcal{P}^1 c_{11}$$

in $H^{27}(E_1)$ our proposition is established.

E_2 of the Massey-Peterson spectral sequence for F_4 at prime 3

$H^*(F_4 ; Z_3) = \Lambda(x_3, \mathcal{P}^1 x_3) \otimes Z_3[y]/(y^3) \otimes \Lambda(x_{11}, \mathcal{P}^1 x_{11})$, $y = \beta \mathcal{P}^1 x_3$

Vertical lines represent multiplication by 3. Elements algebraically coming from $B_5(3)$ are represented by •, from K, by x. Encircled elements, e.g. ⊙, represent infinite towers. The non-zero d_2 is represented and $E_3 = E_\infty$ in the displayed range.

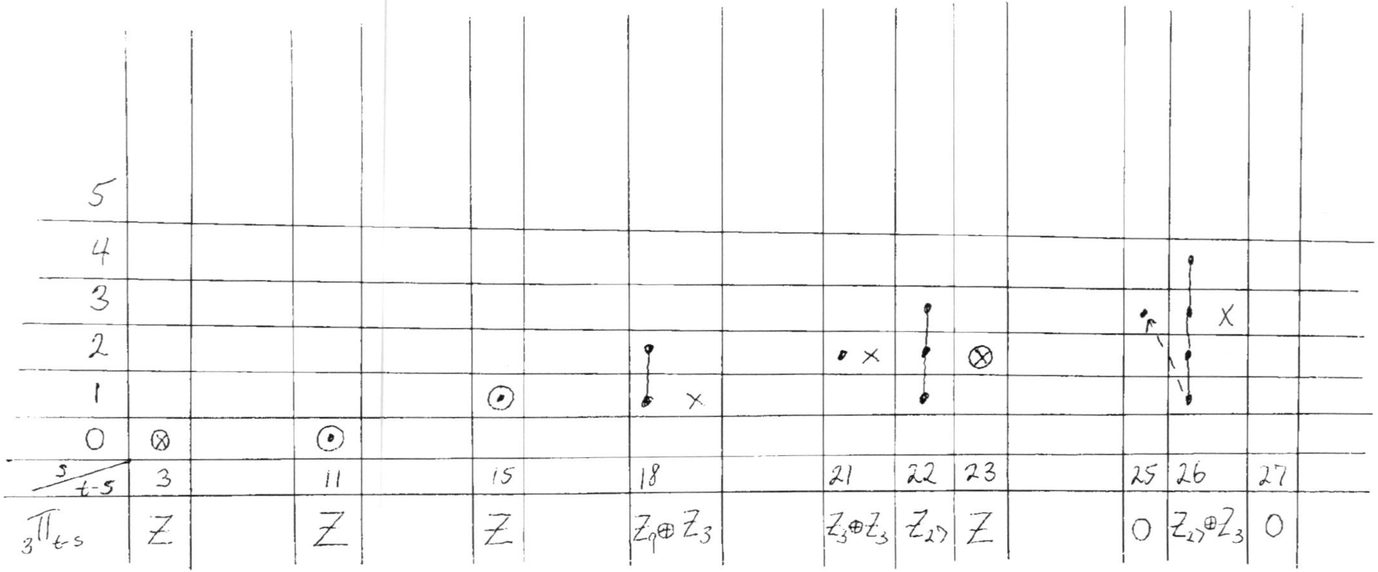

References

[1] W. D. Barcus, On a theorem of Massey and Peterson, Quart. J. Math. 19 (1968) 33-41.

[2] A. K. Bousfield and D. M. Kan, The homotopy spectral sequence of a space with coefficients in a ring, Topology 11 (1972) 79-106.

[3] W. Browder, Higher Torsion in H-spaces, Trans. Amer. Math. Soc. 108 (1963) 353-375.

[4] W. Browder, Homology Ring of Groups, Amer. J. Math. 90 (1968) 318-333.

[5] A. L. Liulevicius, The factorization of cyclic reduced powers by secondary cohomology operations, Amer. Math. Soc. Memoirs 42 (1962).

[6] W. S. Massey and F. P. Peterson, On the mod 2 cohomology structure of certain fibre spaces, Amer. Math. Soc. Memoirs 74 (1967).

[7] L. Smith, Hopf fibration towers and the unstable Adams spectral sequence, Applications of Categorical Algebra, Proc. Sym. Pure. Math. (1970).

[8] A. Zabrodsky, Implication in the cohomology of H-spaces, Ill. J. Math. 14 (1970) 363-375.

[9] Problems in differential and algebraic topology. Seattle conference (1963), R. Lashof editor Ann. of Math. 81 (1965) 565-591.

University of Rochester
Rochester, New York

ON DIRECT LIMITS OF NILPOTENT GROUPS

by

Peter Hilton

Fellow, Battelle Seattle Research Center

Beaumont University Professor, Case Western Reserve University

0. INTRODUCTION

Urs Stammbach [7; p. 170] was the first to point out that the localization theory of nilpotent groups, described in [3], could be extended to direct limits of nilpotent groups. The basic criterion for detecting the P-localizing map $e: G \to G_P$, where P is a family of primes, remains the same, namely, G_P is P-local and e is a P-bijection (called a P-isomorphism in [3]). Stammbach did not mention that a direct limit of nilpotent groups is just a group all of whose finitely generated subgroups are nilpotent, that is, a locally nilpotent group. However, to avoid confusion in the use of the term 'local', we call such groups ℓ-nilpotent and use the symbol $\mathbb{LN}$ for the category of ℓ-nilpotent groups. Then $\mathbb{LN}$ is closed under subgroups, quotient groups, finite products and, of course, direct limits. It is strictly bigger than the category $\mathbb{N}$ of nilpotent groups, since, for example, the restricted direct product of nilpotent groups of class i, $i = 1, 2, \ldots$, is ℓ-nilpotent but not nilpotent.

This paper is directed to extending the purely algebraic theory of localization of nilpotent groups to the category of ℓ-nilpotent groups. It is hoped to follow it, in a sequel, by a paper studying the localization theory of ℓ-nilpotent actions of ℓ-nilpotent groups on commutative groups, as would be required by the intended application to the study of ℓ-nilpotent spaces in homotopy theory. It seems reasonable to conjecture that this is the effective limit of generality for the kind of theory described in [3, 4, 5, 6]. Indeed, there is already one serious potential difficulty in extending both the algebraic and the topological theories to the ℓ-nilpotent case, which consists of the fact that we have not proved a Stallings-Stammbach theorem for ℓ-nilpotent groups; that is, we do not know whether a homomorphism $\phi: G \to K$ of ℓ-nilpotent groups which induces isomorphisms in homology is necessarily an isomorphism.

We open Section 1 by establishing that we may use direct limit arguments to extend the localization functor from $\mathcal{N}$ to $L\mathcal{N}$. Stammbach treats the issue of functoriality somewhat cursorily in [7], confining his argument to the case of direct limits over a common indexing set (and maps respecting the direct systems). We have preferred to base ourselves on a theorem in [2] which turns out to be tailor-made for this particular application. The rest of the section extends the basic results of [3] to $L\mathcal{N}$; we do not trouble to mention those generalizations which are completely automatic.

In section 2 we discuss subgroup theorems. We can dispose immediately of the first main theorem which asserts that if G is ℓ-nilpotent and H, K are subgroups of G, then

$$(H \cap K)_P = H_P \cap K_P, \quad (HK)_P = H_P K_P, \quad [H,K]_P = [H_P, K_P].$$

For the corresponding theorems for nilpotent groups were proved in [4,5], and the extension from $\mathcal{N}$ to $L\mathcal{N}$ is simply achieved by an obvious limit argument. However, we then consider the more delicate question (even in the category $\mathcal{N}$ this question is delicate!) as to when, given a group G, a subgroup H and a P-local ℓ-nilpotent subgroup K, we can infer that $[H,K]$ is P-local. In our discussion of this question we do not insist that G itself be ℓ-nilpotent, though this is, of course, an important special case. We would like to prove results in which we replace the assumption of P-locality (in both hypothesis and conclusion) by the assumption of P'-divisibility[1] (where P' is the complementary set of primes to P); however, although we have such theorems in the nilpotent case, we confine ourselves here to the assumption of P-locality. (Notice that it follows from Theorem 5.2 of [3], extended to ℓ-nilpotent groups, that an ℓ-nilpotent group is P-local if and only if it is P'-divisible and P'-torsion free.)

[1] We sometimes use P' for the (multiplicative) semigroup generated by the primes in P'.

1. LOCALIZING ℓ-NILPOTENT GROUPS

Let $L: \mathcal{N} \to \mathcal{N}_P$ be the P-localizing functor with natural transformation $e: G \to G_P = LG$, $G \in \mathcal{N}$. We will apply Theorem 3.18 of [2] to show that L extends to a unique functor $L_1: L\mathcal{N} \to L\mathcal{N}_P$ (where $L\mathcal{N}_P$ is the category of P-local ℓ-nilpotent groups) such that

(1.1) $$L_1 \underrightarrow{\lim} = \underrightarrow{\lim}\, L.$$

That is, if $G \in L\mathcal{N}$, the category of ℓ-nilpotent groups, and $G = \underrightarrow{\lim}\, G_i$, $G_i \in \mathcal{N}$, then we set

$$L_1 G = \underrightarrow{\lim}\, LG_i .$$

It is easy to see that a direct limit of P-local groups is P-local, so that $\underrightarrow{\lim}\, LG_i \in L\mathcal{N}_P$. Then the justification for adopting (1.1) as a definition of localization in $L\mathcal{N}$ rests on the verification of the hypotheses of Theorem 3.18 of [2].

The first condition to be verified is Hypothesis 3.1 of [2]. This asserts that the pull-back, in the category of groups $\mathcal{G}$, of the diagram

$$\begin{array}{ccc} & & G_o \\ & & \downarrow \\ G'_o & \longrightarrow & G_1 \end{array} \qquad G_o,\ G'_o \in \mathcal{N},\ G_1 \in L\mathcal{N},$$

should belong to $\mathcal{N}$. This, however, is clear (even without the restriction on G_1), since the pull-back is a subgroup of $G_o \times G'_o$. The second condition to be verified (which is evidently necessary) reduces to the following proposition.

<u>Proposition 1.1</u>. *Let $G_i, H_i \in \mathcal{N}$, let $G = \underrightarrow{\lim}\, G_i$, $H = \underrightarrow{\lim}\, H_i$, $G_P = \underrightarrow{\lim}\, G_{iP}$, $H_P = \underrightarrow{\lim}\, H_{iP}$, and let $\phi_i: G_i \to H_i$ be a map of direct systems, localizing to $\phi_{iP}: G_{iP} \to H_{iP}$. Then if ϕ_i, ϕ_{iP} induce*

$$\phi: G \to H,\ \phi_P: G_P \to H_P,$$

respectively, and if ϕ is an isomorphism, so is ϕ_P.

We prove this by a series of lemmas.

<u>Lemma 1.2</u>. *Let $G = \underrightarrow{\lim}\, G_i$, $G_P = \underrightarrow{\lim}\, G_{iP}$. Then the localizing maps $e_i: G_i \to G_{iP}$ induce $e: G \to G_P$, and e is P-bijective.*

<u>Proof</u>. It is easy to show that a direct limit of P-bijections is a P-bijection (see Proposition 6.3 of [7]).

Lemma 1.3. *Given* $A \xrightarrow{\alpha} B \xrightarrow{\beta} C$ *in* G, *then if* $\beta\alpha$ *is P-surjective,* β *is P-surjective; and if* $\beta\alpha$ *is P-injective and* α *is P-surjective,* β *is P-injective.*

Proof. The argument given in Lemma 4.10 of [3] holds in this generality.

Lemma 1.4. *A P-bijection between P-local groups is an isomorphism.*

Proof. Again, we refer to the proof of Lemma 2.8 of [3].

We are now ready to prove Proposition 1.1. For the data give rise to a commutative diagram

$$(1.2)\qquad \begin{array}{ccc} G & \xrightarrow{\phi} & H \\ \downarrow e & & \downarrow e \\ G_P & \xrightarrow{\phi_P} & H_P \end{array}$$

Since ϕ is an isomorphism, it follows from Lemma 1.2 that $\phi_P e = e\phi$ is P-bijective. Since e is P-bijective, it then follows from Lemma 1.3 that ϕ_P is P-bijective. Thus Lemma 1.4 implies that ϕ_P is an isomorphism.

Corollary 1.5. $L\colon N \to N_P$ *extends to* $L_1\colon LN \to LN_P$ *by the rule (1.1). Then* e *extends to a natural transformation from the identity to* EL_1, *where* $E\colon LN_P \subseteq LN$, *and* $e\colon G \to G_P$ *has the universal property.*

Proof. We invoke Theorem 3.18 of [2]. The extension of e to a natural transformation on LN, is attested by (1.2) (where we no longer suppose ϕ an isomorphism); for it follows from the proof of Theorem 3.18 of [2], or it may be proved directly, that, given $\phi\colon G \to H$ in LN, we may find direct systems of nilpotent groups

$$\varinjlim G_i = G, \ \varinjlim H_i = H,$$

over the same directed set, and homomorphisms $\phi_i\colon G_i \to H_i$, of these direct systems, such that $\varinjlim \phi_i = \phi$. It remains to prove the universal property. Given $\phi\colon G \to H$ in LN, with H P-local, we localize to obtain (1.2). However $e\colon H \to H_P$ is an isomorphism by Lemmas 1.2 and 1.4, so that $\psi = e^{-1}\phi_P\colon G_P \to H$ has the property $\psi e = \phi$. This last equation determines ψ, since e is P-surjective and H is P-local.

We may now proceed to generalize to LN the results of [3] with the exception of Theorem 5.7 of [3] and its consequences. For Theorem 5.7 of [3] requires the Stallings-Stammbach Theorem which is unproved for ℓ-nilpotent groups.

We list below only those results which require additional comment.

Proposition 1.6. *L_1: LN $\to$ LN_P is exact.*

Proof. We may either appeal to categorical arguments or reproduce the proof of Proposition 4.6 of [3], which only requires that e be P-bijective.

Proposition 1.7. *Let $G' \rightarrowtail G \twoheadrightarrow G''$ be a short exact sequence in LN. Then if any two of G', G, G'' are P-local, so is the third.*

Proof. We localize and apply Proposition 1.6.

Proposition 1.8. *Let $G' \rightarrowtail G \twoheadrightarrow G''$ be a short exact sequence in LN in which G' is P-local, G is P'-torsion free, and G'' has p^{th} roots, $p \in P'$. Then G and G'' are P-local.*

Proof. We prove that G'' is P'-torsion free, so that G'' is P-local. Let $y \in G''$, $y^n = 1$, $n \in P'$, and $x \mapsto y$, $x \in G$. Then $x^n \in G'$, say $x^n = z$, $z \in G'$. Since G' is P-local, $z = t^n$, $t \in G'$, so $x^n = t^n$. Since G is P'-torsion free it follows (see Theorem 5.2 of [3]) that $x = t$. Thus $y = 1$ and G'' is P'-torsion free. It follows that G'' is P-local and hence, by Proposition 1.7, G is P-local.

The complement to Lemma 1.3 is

Proposition 1.9. *Given $A \xrightarrow{\alpha} B \xrightarrow{\beta} C$ in LN, then if $\beta\alpha$ is P-injective, α is P-injective; and if $\beta\alpha$ is P-surjective, and β is P-injective, α is P-surjective.*

Proof. The first assertion is trivial (and holds for any groups). The second requires an analog of Corollary 6.2 of [3], which we now state.

Lemma 1.10. *Let $G \in$ LN and let $x, y \in G$ with $y^n = 1$. Then there exists c such that $(xy)^{n^c} = x^{n^c}$.*

Proof. Let G_0 be the subgroup of G generated by x, y. Then G_0 is nilpotent and we may assume nil $G_0 \leq c$. The reasoning of Theorem 6.1 and Corollary 6.2 of [3] then shows that $(xy)^{n^c} = x^{n^c}$.

We return to the proof of the second part of Proposition 1.9. If $b \in B$, then, since $\beta\alpha$ is P-surjective, $\beta b^m = \beta\alpha a$ for some $a \in A$, $m \in P'$. Since β is P-injective, $b^m = (\alpha a)u$, where $u \in B$, $u^n = 1$ for some $n \in P'$. Since B is

ℓ-nilpotent, it follows from Lemma 1.10 that there exists c such that $b^{mn^c} = \alpha a^{n^c}$. But $mn^c \in P'$, so α is P-surjective.

Corollary 1.11. *Let* $\phi: G \to H$ *in* LN. *Then* ϕ *is P-bijective if and only if* ϕ_P *is an isomorphism.*

We have remarked that Theorem 5.7 of [3] may not be generalized immediately to LN. However, one half of the theorem goes over immediately (see Proposition 6.1 of [7]).

Proposition 1.12. *Let* $G \in$ LN. *Then* $H_*(e): H_*(G) \to H_*(G_P)$ *P-localizes.*

Proof. Homology and localization commute with direct limits.

2. SUBGROUP THEOREMS

Simply by passing to direct limits (over finitely generated subgroups) we immediately obtain the following generalization of Theorem 1.2 of [5].

Theorem 2.1. *Let* G *be* ℓ*-nilpotent,* H, K *subgroups of* G. *Then*

(i) $(H \cap K)_P = H_P \cap K_P$;

(ii) $(HK)_P = H_P K_P$;

(iii) $[H,K]_P = [H_P, K_P]$.

Our object is now to prove the generalization of Theorem 5.3 of [5] which does not permit such an easy limiting argument. Thus we will prove

Theorem 2.2. *Let* K *be a P-local,* ℓ*-nilpotent normal subgroup of the group* G *and let* $H \subseteq G$. *Then* $[H,K]$ *is P-local.*

We will also prove a variant of this theorem, for which we need a new concept. Let L be a subgroup of G. Then we say that L is *ℓ-normal* in G, if we may write L, G as unions over the same directed set,

$$G = \cup G_i,\ L = \cup L_i,$$

such that L_i is a nilpotent normal subgroup of G_i.

We note the following two special cases of this concept. Let G be ℓ-nilpotent and let L be a normal subgroup of G. Then we have $G = \cup G_i$, where each G_i is nilpotent, and $L = \cup L_i$, where $L_i = L \cap G_i$ is a normal subgroup of G_i

(and hence nilpotent). Second, let L be the (directed) union of nilpotent normal subgroups of G. Then we may take $G_i = G$ for each i, and L is ℓ-normal in G. Notice that an ℓ-normal subgroup is automatically an ℓ-nilpotent normal subgroup.

We prove

Theorem 2.3. Let K be a P-local subgroup of a P'-torsion free, ℓ-normal subgroup L of the group G and let $H \subseteq G$. Then $[H,K]$ is P-local.

We first prove a special case of Theorem 2.3. It is, in fact, the first case given above of an ℓ-normal subgroup.

Proposition 2.4. Let K be a P-local subgroup of a P'-torsion free, normal subgroup L of an ℓ-nilpotent group G. Then $[H,K]$ is P-local.

Proof. Let H_o, K_o be arbitrary finitely-generated subgroups of H, K and let $G_o = H_oK_o$. Then G_o is nilpotent. Moreover $K_o \subseteq K_{oP} \subseteq K$; we claim that $[H_o,K_{oP}] \subseteq G_{oP}$. For $[H_o,K_{oP}]$, as a subgroup of L, is P'-torsion free, and, by Theorem 2.1 (iii),

$$G_{oP} \supseteq [H_o,K_o]_P = [H_{oP},K_{oP}] = [H_o,K_{oP}]_P \supseteq [H_o,K_{oP}].$$

Thus $[H_o,K_{oP}]$ is nilpotent, say nil $[H_o,K_{oP}] \leq c$.

Let $R = [[H_o,K_{oP}],K_{oP}]$. Then R is P-local by Theorem 1.3 of [5], and $R \subseteq [H_o,K_{oP}]$ (see (1.5) of [5]). Now let $f = f(p,c)$ be the Blackburn function (see [1]), such that, if nil $A \leq c$ and $a \in A$ is product of p^{m+f}-powers of elements of A, then a is a p^m-power. Set $n = p^{f+1}$, $p \in P'$, and let $x \in H_o$, $y \in K_{oP}$. Then $y = z^n$, $z \in K_{oP}$, so that

$$[x,y] = [x,z^n] = [x,z]^n u, \; u \in R.$$

Thus $u = v^n$, $v \in R \subseteq [H_o,K_{oP}]$, and $[x,y] = [x,z]^n v^n$. It follows that every element of $[H_o,K_{oP}]$ is a product of nth powers, and hence is a pth power.

Now let $b \in [H,K]$. Then there exist finitely-generated subgroups H_o, K_o of H, K, such that

$$b \in [H_o,K_o] \subseteq [H_o,K_{oP}] \subseteq [H,K].$$

It follows that b has a pth root, $p \in P'$, in $[H_o,K_{oP}]$ and hence in $[H,K]$. Since $[H,K]$, as a subgroup of L, is P'-torsion free, the proposition is proved.

Note that we have proved that $[H_o,K_{oP}]$ is P-local; it thus follows that

$$[H_o,K_o]_P = [H_o,K_{oP}] = [H_{oP},K_{oP}]. \tag{2.1}$$

Proposition 2.5. Under the hypotheses of Theorem 2.2 or Theorem 2.3, $[[H,K],K]$ is P-local.

Proof. To obtain the conclusion under the hypotheses of Theorem 2.2, we apply Proposition 2.4 with H, K, L, G replaced by $[H,K]$, K, K, K respectively. To obtain the conclusion under the hypotheses of Theorem 2.3, we apply Proposition 2.4 with H, K, L, G replaced by $[H,K]$, K, L, L.

Proof of Theorem 2.2. We consider the short exact sequence

$$[[H,K],K] \rightarrowtail [H,K] \twoheadrightarrow [H,K]/[[H,K],K]. \tag{2.2}$$

Since $[H,K] \subseteq K$ it follows that $[H,K]/[[H,K],K]$ is commutative. Moreover, since

$$[a,b^n] \equiv [a,b]^n \bmod [[H,K],K],\ a \in H,\ b \in K,$$

and since K is P-local, we readily infer that $[H,K]/[[H,K],K]$ has p^{th} roots, $p \in P'$. Thus Theorem 2.2 is proved by an application of Proposition 1.8.

Proof of Theorem 2.3. We again base ourselves on (2.2) but we must first establish that $[[H,K],K]$ is normal in $[H,K]$. However, for c, $d \in [H,K]$, $k \in K$, we have

$$d^{-1}c^{-1}k^{-1}ckd = d^{-1}c^{-1}k^{-1}cdkk^{-1}d^{-1}kd = [cd,k][d,k]^{-1} \in [[H,K],K],$$

as required.

Now let $G = \cup G_i$, $L = \cup L_i$, where L_i is a nilpotent normal subgroup of G_i. Set $K_i = K \cap L_i$, $H_i = H \cap G_i$. Then $K = \cup K_i$, $H = \cup H_i$ and

$$[H,K]/[[H,K],K] = \varinjlim [H_i,K_i]/[[H_i,K_i],K_i]. \tag{2.3}$$

This is clear since, for any elements of $[H,K]$ there exist finitely-generated subgroups $H_o \subseteq H$, $K_o \subseteq K$ such that $x \in [H_o,K_o]$, and then there exist i_1, i_2 such that $H_o \subseteq H_{i_1}$, $K_o \subseteq K_{i_2}$. Taking $i \geq i_1$, $i \geq i_2$, we have $H_o \subseteq H_i$, $K_o \subseteq K_i$. Similarly we find that $\varinjlim [[H_i,K_i],K_i] = [[H,K],K]$, so that (2.3) follows. However, since K is P-local we also have $K = \bigcup K_{iP}$, so we may modify (2.3) to

$$(2.4) \qquad [H,K]/[[H,K],K] = \varinjlim [H_i,K_{iP}]/[[H_i,K_{iP}],K_{iP}]$$

We want to show that $[H,K]/[[H,K],K]$ has p^{th} roots, $p \in P'$; for then we will complete the proof by appeal to Proposition 1.8 (note that $[H,K]$, as a subgroup of L, is certainly ℓ-nilpotent). Thus, by (2.4), we must show that $[H_i,K_{iP}]/[[H_i,K_{iP}],K_{iP}]$ has p^{th} roots, $p \in P'$. Now K_i is a subgroup of L_i which is normal in G_i. Thus, for each $x \in H_i$, $x^{-1}K_i x \subseteq L_i$. Now $K_i \subseteq K_{iP} \subseteq K \subseteq L \subseteq G$, so we may form $x^{-1}K_{iP}x$, which is obviously the localization of $x^{-1}K_i x$. It follows that $x^{-1}K_{iP}x \subseteq L_{iP}$, $x \in H_i$, so that

$$[H_i,K_{iP}] \subseteq L_{iP}.$$

We infer that $[H_i,K_{iP}]$ is nilpotent and our standard argument, exploiting the Blackburn function, now shows that $[H_i,K_{iP}]/[[H_i,K_{iP}],K_{iP}]$ has p^{th} roots, $p \in P'$. This completes the proof of Theorem 2.3.

Our final remark concerns the generalization of (2.1).

Proposition 2.6. *Assume the hypotheses of Proposition 2.4 and let* S *be a subgroup of* K. *Then*

$$[H,S]_P = [H,S_P].$$

Proof. We have $S \subseteq S_P \subseteq K$. Thus the hypotheses of Proposition 2.4 apply with K replaced by S_P, so that $[H,S_P]$ is P-local. On the other hand, by Theorem 2.1 (iii),

$$[H,S_P]_P = [H_P,S_P] = [H,S]_P,$$

so that $[H,S_P] = [H,S]_P$, as required.

It is reasonable to conjecture that the conclusion of Proposition 2.6 remains valid under the hypotheses of Theorem 2.3.

BIBLIOGRAPHY

1. N. Blackburn, Conjugacy in nilpotent groups, Proc. Amer. Math. Soc. 16 (1965), 143-148.
2. P. Hilton, On the category of direct systems and functors on groups, Journ. Pure and App. Alg. 1 (1971), 1-26.
3. P. Hilton, Localization and cohomology of nilpotent groups, Math. Zeits. 132 (1973), 263-286.
4. P. Hilton, Remarks on the localization of nilpotent groups, Comm. Pure and App. Math. (1974).
5. P. Hilton, Nilpotent actions on nilpotent groups, Proc. Austr. Summer Institute (1974).
6. P. Hilton, G. Mislin and J. Roitberg, Localization of nilpotent groups and spaces (to appear).
7. U. Stammbach, <u>Homology in Group Theory</u>, Lecture Notes in Mathematics 359, Springer Verlag (1973).

ARITHMETIC K-THEORY

R. Hoobler and D. L. Rector

Rice University

1. INTRODUCTION

The aim of our present joint work is to make accessible to algebraic topologists some of the techniques of étale homotopy and cohomology theory. In this note we outline the techniques needed to give the "right" proof of the following theorem of Tørnehave.

Recall Quillen's definition of the algebraic K-groups of a field F with discrete topology. Consider

$$BG\ell(F) = \varinjlim BG\ell(n,F).$$

One may attach cells to this space to obtain a simple space $BG\ell^+(F)$ with the same cohomology. Then for $i > o$,

$$K_iF = \pi_i\ BG\ell^+(F).$$

One way to construct this space is to note that the space

$$B_* = \coprod_n BG\ell(n,F)$$

is a free simplicial monoid with product given by the obvious inclusions

$$BG\ell(n,F) \times BG\ell(m,F) \to BG\ell(n+m,F)$$

"representing Whitney sum". If UB_* is the simplicial group obtained by group completing B_* dimensionwise, then

$$UB_* = \mathbb{Z} \times BG\ell^+(F)$$

1. talk presented by D. L. Rector

Using the techniques of Segal [15], et.al. , $\mathbb{Z} \times BG\ell^{+}(F)$ can be made part of a connected Ω-spectrum, $\underline{\underline{KF}}$, giving rise to a cohomology theory with

$$KF^{i}(pt) = K_{-i}(F) .$$

Now for $F = \mathbb{F}_q$, a finite field of q-elements, Quillen constructed, using modular character theory, a map

$$Q: BG\ell^{+}(\mathbb{F}_q) \to BU$$

so that

$$BG\ell^{+}(\mathbb{F}_q) \to BU \xrightarrow{\Psi^q - 1} BU$$

is a quasi fibration [T4] .

Let $\underline{\underline{KC}}$ be the spectrum of connective complex K-theory, $\underline{\underline{KC}}[\frac{1}{p}]$ that spectrum localized away from p . A theorem of Tørnehave [T6] is

Theorem 1.1. *There is a multiplicative map of spectra*

$$\underline{\underline{KF}}_q \to \underline{\underline{KC}}[\frac{1}{p}],$$

extending the Quillen map, and such that

$$\underline{\underline{KF}}_q \to \underline{\underline{KC}}[\frac{1}{p}] \xrightarrow{\Psi^q - 1} \underline{\underline{KC}}[\frac{1}{p}]_0$$

is a fibration of spectra, where $\underline{\underline{KC}}[\frac{1}{p}]_0$ *is the 0-connected cover of* $\underline{\underline{KC}}$.

Tørnehave's proof is a calculation, using modular character theory, of the appropriate higher homotopies. We would like to show how a more natural proof of this theorem can be given once the appropriate general tools are available and to outline how those tools may be obtained.

2. OUTLINE OF PROOF

I. The spectra $\underline{\underline{KF}}$ and $\underline{\underline{KC}}$ may be obtained from

1) the spaces $B_i = BG\ell(n,F)$ or $BG\ell(n,\mathbb{C})$
2) the "Whitney sum" maps

$$B_i \times B_j \to B_{i+j}$$

3) the compatible action of the permutation groups Σ_n on B_n.

To construct maps of spectra, it suffices to give maps $BG\ell(n,F) \to BG\ell(n,\mathbb{C})$ preserving the extra structure. For example, if F is a subfield of $\mathbb{C}$ with F having the discrete topology, then the inclusion $F \to \mathbb{C}$ is continuous and induces the appropriate maps.

II. If G is an algebraic group over a field F, then G is a variety and has an étale topology. The étale topology of G captures a profinite homotopy type for G; so we may hope to define in some similar way a classifying space BG_{et} for G. We show how this may be done in §3.

III. The maps $G\ell(n,F) \times G\ell(m,F) \to G\ell(n+m,F)$ and the action of Σ_n on $G\ell(n,F)$ are all algebraic, and thus induce actions on $BG\ell(n,F)_{et}$. We may thus hope to construct a spectrum $\underline{\underline{K_{et}F}}$. In §4, we outline how this may be done using a generalization of the techniques of Segal-Anderson.

IV. General principles of étale theory then indicate there will be homotopy equivalences

$$\widehat{\underline{\underline{K_{et}\overline{\mathbb{F}}_q}}[\tfrac{1}{p}]} \xleftarrow[\sim]{\hat{p}} \xi \xrightarrow{\sim} \widehat{\underline{\underline{K_{et}\mathbb{C}}}} \xleftarrow{\sim} \xi \xrightarrow{\sim} \widehat{\underline{\underline{KC}}},$$

where $\widehat{\ }$ denotes profinite completion, $\xrightarrow{\hat{p}}$ denotes homotopy equivalence away from $p = \operatorname{char} \mathbb{F}_q$, $\overline{\mathbb{F}}_q$ is the algebraic closure of $\mathbb{F}_q$, and the ξ are intermediate objects we need not discuss (some of which have been left out).

V. Since "the discrete topology on $\overline{\mathbb{F}}_q$ is finer than the étale topology", the inclusion $\mathbb{F}_q \longrightarrow (\overline{\mathbb{F}}_q)_{et}$ induces maps

$$BG\ell(n,\mathbb{F}_q) \longrightarrow BG\ell(n,\overline{\mathbb{F}}_q)_{et}$$

preserving the extra structure and therefore induces a map $\underline{\underline{K\mathbb{F}}}_q \to \underline{\underline{K_{et}\overline{\mathbb{F}}_q}}$ (These statements actually require a bit of interpretation, but will serve to indicate the ideas). Now there is a homotopy equivalence

$$\widehat{\underline{\underline{K_{et}\overline{\mathbb{F}}_q}}}[\tfrac{1}{p}] \xrightarrow{\simeq} \widehat{\underline{\underline{KC}}}[\tfrac{1}{p}] \;;$$

so we have a map $\underline{\underline{K\mathbb{F}}}_q \to \widehat{\underline{\underline{KC}}}[\frac{1}{p}]$. But $\pi_i\underline{\underline{K\mathbb{F}}}_q$, $i > 0$, are finite so there are no obstructions to lifting this map to a map $Q: \underline{\underline{K\mathbb{F}}}_q \to \underline{\underline{KC}}[\frac{1}{p}]$.

VI. Since the $G\ell(n,\overline{\mathbb{F}}_q)$ are defined by equations with coefficients in $\mathbb{F}_q$, the Galois group of $\overline{\mathbb{F}}_q$ over $\mathbb{F}_q$ acts on the $BG\ell(n,\overline{\mathbb{F}}_q)_{et}$. Quillen has shown [T3], that the Frobenius φ, $\varphi x = x^q$, $x \in \overline{\mathbb{F}}_q$, induces the Adams operation Ψ^q on $\underline{\underline{K_{et}\overline{\mathbb{F}}_q}}$. But φ has the discrete $G\ell(n,\mathbb{F}_q)$ as its fixed points, so Q maps to the fibre of $(\Psi^q - 1)$. The rest of the proof follows from Quillen's cohomology arguments.

3. ÉTALE HOMOTOPY TYPES

A variety is built up in much the same way as a differentiable (or analytic) manifold. A differentiable (analytic) manifold is covered by open sets each of which is homeomorphic to an open subset of $\mathbb{R}^n(\mathbb{C}^n)$ and which are patched together by differentiable (analytic) maps. If U is an open set of M one may associate to U the ring $\mathcal{O}_{M,U}$ of differentiable (analytic) functions on U with values in $\mathbb{R}(\mathbb{C})$. These rings form a sheaf $\mathcal{O}_M$ whose stalk at $x \in M$, $\mathcal{O}_{M,x}$, is called the ring of <u>germs of differentiable (holomorphic) functions</u> at x. The structure of M is determined by the space M and the sheaf $\mathcal{O}_M$ since <u>a continuous function</u> $f : M \to N$ <u>is differentiable iff it is covered by a map of sheaves</u> $f^* : \mathcal{O}_N \to \mathcal{O}_M$.

Let K be an algebraically closed field. The basic building blocks of varieties over K are the <u>irreducible affine varieties</u> which are the subsets V of K^n such that V is the set of zeros of a set of polynomials $f_1, \ldots f_m$ which generate a prime ideal $\mathfrak{A} \subseteq K[X_1, \ldots, X_n]$. In particular, affine n-space $\mathbb{A}_K^n$ corresponds to the ideal (0) . $\mathbb{A}_K^n$ has the <u>Zariski topology</u> generated by open sets $\mathbb{A}_K^n - V$, V an irreducible affine variety. Such a variety V inherits its Zariski topology from $\mathbb{A}_K^n$. The Zariski topology is a very coarse topology; in particular, every non empty open set is dense.

If V is an irreducible affine variety defined by a prime ideal $\mathfrak{A}$, we may associate to V its <u>affine coordinate ring</u>

$$A = K[X_1, \ldots, X_n] / \mathfrak{A}$$

and its ring of <u>rational functions</u> R_V the quotient field of A. V then may be given a <u>structure sheaf</u> $\mathcal{O}_V$ such that

$$\mathcal{O}_{V,x} = \{f/g \in R \mid g(x) \neq 0\}.$$

These are <u>local rings</u>. Then $\mathcal{O}_{V,V} = A$. A variety is now essentially a space and sheaf of rings covered by open affine varieties. In particular any open set in $\mathbb{A}^n$ is also a variety, and any subvariety V of $\mathbb{A}^n$ is determined by its coordinate ring $\mathcal{O}_{V,V}$.

<u>Examples</u>: Let K^{n^2+1} have coordinates $X_{11}, \ldots, X_{nn}, D$. Then $G\ell(n,K)$ is the subvariety defined by the equation

$$(1 - \det \cdot D) = 0$$

where $\det$ is the determinant of the matrix (X_{ij}). Its coordinate ring is

$$K[X_{11},\ldots,X_{nn}][\det^{-1}]$$

Similarly $SL(n,K) \subseteq K^{n^2}$ is the zeros of $\det - 1$ and has coordinate ring

$$K[X_{11},\ldots,X_{nn}] \;/\; (\det - 1)\;.$$

It is worth noting that the coordinate rings of algebraic groups are Hopf algebras. For $n=1$, $G\ell(1,K) \overset{df}{=} G_m = \mathbb{A}^1_K - \{0\}$ and has coordinate ring $K[T,T^{-1}]$.

The notion of variety may be extended to non algebraically closed fields by the notion of <u>scheme</u> (see general references). We will not need to discuss the technicalities except to remark that "points" of $V \subseteq K^n$ now become the <u>maximal ideals</u> of $\mathcal{O}_{V,V}$. (The intuitive reason for this may be seen by looking at the Hilbert Nullstellensatz).

If X is an algebraic variety over $\mathbb{C}$, a theorem essentially going back to Riemann asserts that any finite covering space of X has a unique structure as an algebraic variety. This suggests that the profinite completion of $\pi_1 X$ might be recovered by purely algebraic techniques. This was achieved by Grothendieck with his definition of the étale topology of a scheme [G2]. A possibly more readable approach is given in [G1]. For the constructions in §2, except for some of the intermediate objects in IV, we only need to discuss smooth varieties over a field. This will simplify our definitions. Good general references are [G5] and [G6], the former being more elementary.

Now if $f: X \to Y$ is a map of differentiable manifolds, the implicit function theorem gives a criteron for a map to be a local diffeomorphism. This criterion can be mimicked in the algebraic category to give the notion of an <u>étale morphism</u>. But the Zariski topology is too coarse for the implicit function theorem to be true in its usual form. The definition of étale for smooth varieties may be given in terms of power series at a point. Since our fields may not be algebraically closed we will need the notion of geometric point.

Definition: 1) A geometric point $i_{\bar{y}}:\bar{y} \to Y$ of an algebraic variety over a field K is a pair (y,φ_y) where y is a point of Y and $\varphi_y:\mathcal{O}_{Y,y} \to K_s$ is a K-algebra homomorphism from $\mathcal{O}_{Y,y}$, the local ring of $y \in Y$, into K_s, a separable closure of K. Given a geometric point (y,φ_y), let $\bar{\varphi}_y:K_s \otimes_K \mathcal{O}_{Y,y} \to K_s$ be the natural extension of φ_y. If $m_{\bar{y}}$ is the kernel of $\bar{\varphi}_y$, let $\mathcal{O}_{Y,\bar{y}} = [K_s \otimes_K \mathcal{O}_{Y,y}]_{m_{\bar{y}}}$. It is called the local ring of the geometric point $\bar{y}$.

2) An algebraic variety X is smooth if for any geometric point (y,φ_y), $\hat{\mathcal{O}}_{Y,\bar{y}} \simeq K_s[[T_1,\ldots,T_r]]$ where $r = \dim X$ and the $\wedge$ denotes completion with respect to the topology defined by powers of the unique maximal ideal.

3) Let X and Y be algebraic varieties over a field K with Y smooth. A map $\pi:X\to Y$ of varieties is étale at $x \in X$ if the map $\pi^*:\hat{\mathcal{O}}_{Y,\overline{\pi(x)}} \to \hat{\mathcal{O}}_{X,\bar{x}}$ is an isomorphism for any geometric point of the form (x,φ_x). π is étale if it is étale at all points of X.

Example: Let X and Y be $\mathbb{A}^1-\{0\}$, the affine line over Q with the origin deleted. Define $\pi:X\to Y$ by $\pi(x)=x^n$. Then $\pi^*:K[T_Y,T_Y^{-1}] \to K[T_X,T_X^{-1}]$ is given by $\pi^*(T_Y) = T_X^n$ where T_X,T_Y are the coordinate functions of X and Y respectively. There are n distinct geometric points of X over any geometric point $\bar{y}$ of Y which are gotten by $\varphi_i(T_X) = \zeta^i(\varphi_y(T_Y))^{1/n}$, ζ a primitive nth root of unity. Moreover the points x of X corresponding to these geometric points have as their local rings $K[X]_m$ where m is one of the maximal ideals generated by the mth cyclotomic polynomial, for $m|n$. Consequently $\bar{\mathbb{Q}}\otimes_{\mathbb{Q}}\mathcal{O}_{X,x}$ is a semi-local ring whose maximal ideals correspond to the monic factors of the cyclotomic polynomial over $\bar{\mathbb{Q}}$.
The equation $T_X^n - \alpha = \Pi(T_X - \zeta^i\alpha^{1/n})$ shows that π is étale.

Similar examples can be constructed over $\mathbb{F}_q$ as long as <u>n</u> <u>and</u> <u>q</u> <u>are relatively prime</u>.

Étale morphisms have the following properties:

1) If $f:X\to Y$, $g:Y\to Z$ are étale, so is $gf:X\to Z$.

2) If $f:X\to Y$ is étale, $Z\to Y$ arbitrary, $X\times_Y Z\to Z$ is étale.

3) If $f_1:X_1\to Y_1$, and $f_2:X_2\to Y_2$ are étale, so is $f_1\times f_2$.

4) If Y is smooth and $f:X\to Y$ is étale, then X is smooth.

5) If $f:X\to Y$ is étale, $f(X)$ is a Zariski open set of Y .

To construct the étale homotopy type of a variety, we would like to mimic the Cech construction. Note that if U and V are open sets of a space X, then $U\cap V = U\times_X V$. We extend the notion of covering to a covering by a collection of étale maps. Note that we are essentially "forcing the implicit function theorem" in its classical form.

<u>Definition</u> 3.1. Let $Pt(X)$ denote the set of geometric points of X . The catagory of <u>pointed coverings</u> of X, Cov.(X), is defined as follows.

A pointed covering is a collection

$$\{(\pi_y:U_y\to X,i_y)\}_{y\in Pt(X)}\ ,$$

where π_y is étale, and i_y is a geometric point of U_y such that $\pi_y(i_y) = y$. A map $\Phi:\{(U_y\to X,i_y)\}\to\{(V_y\to X,j_y)\}$ is a collection of maps $\varphi_y:U_y\to V_y$ over X which preserve geometric points. The category Cov.(X) has fibred products and is a good category for limits (pseudo filtering).

We can use the standard Cech complex to define the Cech homotopy type of X . For $\mathfrak{U}\in$ Cov.(X), let $C_{\mathfrak{U}}(X)_*$ denote the simplicial variety defining the Cech complex. Then $C_{\mathfrak{U}}(X)_n =$

$$\coprod_{Pt(X)^{n+1}} U_{y_0}\times U_{y_1}\times U_{y_2}\times\cdots\times U_{y_n}$$

and the face and degeneracy operators come from projections omitting a factor and diagonal maps on a given factor.

Definition 3.2. $\Pi_{\mathfrak{U}}(X)$ is the simplicial set $\Pi_0(C_{\mathfrak{U}}(X)*)$ where Π_0 is the connected component functor. If X has a base point, then $\Pi_{\mathfrak{U}}(X)$ also has a base point. $\Pi(X) = \{\Pi_{\mathfrak{U}}(X)\}_{\mathfrak{U} \in \mathrm{Cov}(X)^{op}}$ is the pro-object constructed from all of the pointed coverings of X.

The most important theorem about this construction is essentially due to M. Artin [G2]

Theorem 3.3. *Let X be a smooth algebraic variety over $\mathbb{C}$ which is a (not necessarily closed) subvariety of $\mathbb{P}_n$, A a finite group which is abelian if $n > 1$. Then there is an isomorphism $H^n(X,A) \simeq \varinjlim_{n} H^n(\Pi_{\mathfrak{U}}(X);A)$, where $H^n(X;A)$ is the singular cohomology of X in the classical topology.*

Let G be an algebraic group over a field K, e.g., $G = Gl(n,K)$. Grothendieck has introduced an algebraic analogue of BG, denoted $\mathbb{B}_G$. Its definition essentially forces the universal mapping property for G bundles. He has also given an algebraic definition of its cohomology groups. (For some beautiful applications to topology see [G3] and [G4]). We will recover a homotopy type for $\mathbb{B}_G$ by a more geometric method.

Consider the bar construction $B.G$. $B.G$ is a simplicial variety with

$$(B.G)_n = \prod_{i=1}^{n} G$$

and faces given by multiplication. In defining BG_{et}, one would like to apply the above construction to $B.(G)$, but unfortunately this gives a simplicial pro-space rather than a pro-simplicial space. Consequently we must first extend our notion of pointed coverings to simplicial varieties.

Definition 3.4. Let $X. = (X_n, d_i^n, s_i^n)$ be a simplicial variety.

$$\text{Ob Cov.}(X.) = \{(\mathfrak{U}_n, \partial_i^n, \sigma_i^n) \mid \mathfrak{U}_n \in \text{Cov.}(X_n)\}$$

where $\partial_i^n: \mathfrak{U}_n \to \mathfrak{U}_{n-1}$ and $\sigma_i^n: \mathfrak{U}_n \to \mathfrak{U}_{n+1}$ cover d_i^n and s_i^n respectively. A map between two pointed simplicial coverings $\Phi: (\mathfrak{U}_n, \partial_i^n, \sigma_i^n) \to (\mathcal{V}_n, \bar{\partial}_i^n, \bar{\sigma}_i^n)$ is a tuple of maps $\varphi_n: \mathfrak{U}_n \to \mathcal{V}_n$ of pointed coverings of X_n which commute with the respective face and degeneracy operators.

Elements of Cov.(X.) will in general be denoted by $\mathfrak{U}$ unless there is a possibility of confusion. The functorial properties of pointed simplicial coverings are essentially the same as for pointed coverings although there is an additional layer of commuting diagrams to be checked. Thus Cov.(X.) has finite fibred products and a final object. Moreover there is a forgetful functor F_N: Cov.(X.) $\to$ Cov.(X_N) defined by $F_N((\mathfrak{U}_n, \partial_i^n, \sigma_i^n)) = \mathfrak{U}_N$. The key result which allows us to transform our simplicial pro-space into a pro-simplicial space is the following finality assertion.

Lemma 3.5. Given a simplicial variety $X. = (X_n, d_i^n, s_i^n)$ and a non-negative integer N , there is a functor E_N: Cov.(X_N) $\to$ Cov.(X.) and a natural transformation $\eta: F_N E_N \to 1$.

Corollary: 3.6. The functor F_N is cofinal.

We now define the homotopy type of $\mathbb{B}_G$.

Definition 3.6. Given an affine algebraic group G over a field k as above, $\Pi(\mathbb{B}_G) = \{\Pi_{\mathfrak{U}.}(\mathbb{B}_G)\}_{\mathfrak{U}. \in \text{Cov.}(B.(G))^{op}}$ where $\Pi_{\mathfrak{U}.}(\mathbb{B}_G)$ is the diagonal simplicial set associated to the simplicial space $\Pi_{\mathfrak{U}.}(B.(G))$. Fixing a base point $\bar{y}$ for X determines a base point of $\Pi_{\mathfrak{U}.}(B.(G))$ and so one for $\Pi_{\mathfrak{U}.}(\mathbb{B}_G)$. The main result required to show that this is a reasonable definition is the verification that cohomology with locally constant coefficients in $\mathbb{B}_G$ can be computed from this pro-simplicial set.

Theorem 3.7. *Let* G *be an affine algebraic group defined over a field* k , A *an abelian group*. Then $H^n(\mathbb{B}_G,\underline{A}) \cong \varinjlim_n H^n(\pi_{\mathfrak{U}.}(\mathbb{B}_G);A)$.

Theorem 3.8. *Let* G *be an affine algebraic group over* $\mathbb{C}$, A , *a finite abelian group which is commutative if* $n > 1$. *Then*

$H^n(\mathbb{B}_G,\underline{A}) \simeq H^n(B_G,A)$ *where the latter group is the singular cohomology of the classifying space for* G *over* $\mathbb{C}$.

Remark: It is easy to construct examples of a space X whose étale and singular cohomology with integer coefficients are different. The traditional way of dealing with this problem is to define $H^*_\ell(\mathbb{B}_G) = \varprojlim_n H^*(\mathbb{B}_G,\mathbb{Z}/\ell^n\mathbb{Z})$ if G is over an algebraically closed field. More generally, one of the advantages of the étale cohomology of a variety X defined over k is that the Galois group of k acts naturally on $H^*(X,\underline{A})$. Consequently one introduces ℓ^n th roots of unity as coefficients in order to keep better track of this action. But over an algebraically closed field we are non-canonically reduced to the above group.

4. CONSTRUCTION OF SOME CHAIN FUNCTORS

To construct the homology theories in which we are interested, it will be sufficient to construct a chain functor, [T1], [Anderson, these notes]

$$\Phi : (\text{pointed finite sets}) \to (\text{spaces})$$

such that the natural map

$$\Phi(S \vee T) \to \Phi(S) \times \Phi(T)$$

is a homotopy equivalence. We will generalize the following construction of D. Anderson [T1].

Let P be a permutative category; that is, a small category together with a composition law (functor)

$$+ : P \times P \to P$$

making Map (P) into a monoid and a natural transformation α between $+$ and $+ \circ T$, where $T: P\times P \to P\times P$ sends (f,g) to (g,f), such

1) $\alpha(A,B) = \alpha(B,A)^{-1}$
2) $(\alpha(A,C) + B) \circ (A+\alpha(B,C)) = \alpha(A+B,C)$.

The natural transformation α says that $+$ is "coherently commutative". The category of permutative categories has a product and coproduct. $P \sqcap Q$ is just $P \times Q$, where $(f,g) + (f',g') = (f+f',g+g')$. One has

$$\mathcal{O}b(P \sqcup Q) = \mathcal{O}b(P) * \mathcal{O}b(Q)$$

where $*$ denotes free product as monoids. So the objects of $P \sqcup Q$ can be described uniquely as reduced words. The maps of $P \sqcup Q$ are generated by composition from

1) the maps of P and Q.
2) $+$ being a composition law
3) for each $p \in \mathcal{O}b(P)$, $q \in \mathcal{O}b(Q)$ a formal equivalence

$\alpha(p,q) : p+q \to q+p$.

The natural map $P \sqcup Q \to P \sqcap Q$ has a section $P \sqcap Q \to P \sqcup Q$ sending (p,q) to $p+q$. It is important to us that <u>these maps constitute a natural equivalence of categories</u>.

For P a permutative category, we may now obtain a chain functor as follows. For S a finite pointed set, let

$$P_s = \begin{cases} \text{a copy of } P & \text{if } s \neq * \\ * & , s = * \end{cases}$$

where $*$ denotes above both the base point of a set and the trivial permutative category. Now $\coprod_S P_s$ is functorial in S. So let

$$\Phi(S) = N(\coprod_{S} P_s)$$

where N is the nerve of a category [T5]. That Φ is a chain functor follows immediately from

1) the nerve of a natural equivalence is a homotopy equivalence

2) $N(P_1 \times P_2) = NP_1 \times NP_2$

3) $P_1 \coprod P_2' \to P_1 \Pi P_2$ is a natural equivalence.

The theories $\underline{\underline{KF}}_q$ and $\underline{\underline{KC}}$ can both be obtained from this construction. However, it does not suffice for our purposes, since $BG\ell(n,F)_{et}$ cannot be obtained as the nerve of a convenient category. We will generalize this construction by treating separately the topology of the $BG\ell(n,F)_{et}$ and the topology of the permutative structure.

We may codify the properties of the Whitney sum maps and the action of the Σ_n on the spaces $BG\ell(n,F)_{et}$ as follows. Let Σ be the category such that

1) $\mathcal{O}b\Sigma = \{\underline{\underline{n}} \mid n = 0,1,2,3,\cdots\}$

2) $\mathrm{Hom}(\underline{\underline{n}},\underline{\underline{n}}) = \Sigma_n$

3) $\mathrm{Hom}(\underline{\underline{n}},\underline{\underline{m}}) = \emptyset$, $n \neq m$.

Thus $\Sigma = \coprod_{n \in \mathbb{N}} \Sigma_n$, where the group Σ_n is considered as a category with one object. We may also consider Σ to be the category of finite sets $\underline{\underline{n}}$ where $\underline{\underline{n}} = \{1,2,\cdots,n\}$, and bijections of sets. One then sees immediately that Σ is a permutative category under disjoint union, so that $\underline{\underline{n}}+\underline{\underline{m}} = \underline{\underline{n+m}}$, where $\underline{\underline{n+m}} = \{1,...,n\} \cup \{n+1,...,n+m\}$.

Definition 4.1. If $\underline{\underline{C}}$ is a category with finite products and a zero object $*$, a Σ-structure in $\underline{\underline{C}}$ is a functor $\Gamma:\Sigma \to \underline{\underline{C}}$ such that $\Gamma\,\underline{\underline{0}} = *$ together with a natural transformation $\varphi: \Gamma \times \Gamma \to \Gamma \circ +$, where $(\Gamma \times \Gamma)(p,q) = \Gamma p \times \Gamma q$.

The examples of Σ-structures which interest us here are those when $\underline{\underline{C}}$ is either (groups), (topological groups), or (algebraic groups),

$$\Gamma\underline{\underline{n}} = G\ell(n,F),$$

and φ is given by the Whitney sum maps.

Now if G is a group, let $\tilde{G}$ be the category whose objects are the elements of q and such that there is exactly one map $g \to g'$ for $g,g' \in G$. Then

$$N\tilde{G} = WG ,$$

the acyclic bar construction. Let $\tilde{\Sigma} = \coprod_n \Sigma_n$.

Given $\Gamma : \Sigma \to \underline{\underline{C}}$, we have a functor $\tilde{\Gamma}:\tilde{\Sigma} \to \underline{\underline{C}}$ given by

$$\tilde{\Gamma}\sigma = \Gamma\underline{\underline{n}} , \quad \sigma \in \Sigma_n$$

$$\tilde{\Gamma}(\sigma\to\sigma') = \Gamma(\sigma'\sigma^{-1}) .$$

Note that $\tilde{\Sigma}$ is also a permutative category (i.e., finite <u>ordered</u> sets under disjoint union) and the natural transformation φ induces a natural transformation $\tilde{\varphi}:\tilde{\Gamma}\times\tilde{\Gamma} \to \tilde{\Gamma} \circ +$. Furthermore, $\tilde{\Gamma}$ has a property which we generalize in

<u>Definition</u> 4.2. Let P be a permutative category, $\underline{\underline{C}}$ a category with product and a zero object. A <u>permutative functor</u> is a functor $\Gamma:P \to \underline{\underline{C}}$, $\Gamma * = *$, together with a natural transformation $\tilde{\varphi}:\tilde{\Gamma}\times\tilde{\Gamma} \to \tilde{\Gamma} \circ +$ such that for $p,p' \in \mathcal{O}b(P)$

$$\begin{array}{ccc}
\Gamma_p \times \Gamma_{p'} & \xrightarrow{T} & \Gamma_{p'} \times \Gamma_p \\
\Big\downarrow \varphi(p,p') & & \Big\downarrow \varphi(p',p) \\
\Gamma(p+p') & \xrightarrow{\alpha(p,p')} & \Gamma(p'+p)
\end{array}$$

commutes.

Let $\{P_s\}$, $s \in S$, be a finitely indexed set of permutative categories, $\Gamma_s : P_s \to \underline{\underline{C}}$ permutative functors. Define

$$\Gamma_S : \coprod_S P_s \to \underline{\underline{C}}$$

as follows. Let

$$\coprod_S P_s \xrightarrow{\gamma} \prod_S P_s$$

be the natural functor. If $x = \Sigma p_i$, $p_i \in P_{s_i}$, is a word in $\mathcal{O}b(\coprod P_s)$ then

$$\gamma(x) = \prod_{s \in S} (\sum_{s_i = s} p_i)$$

Put

$$\Gamma_S(x) = (\prod_S \Gamma \circ \gamma)(x)$$

The natural transformation $\varphi_S : \Gamma_S \times \Gamma_S \to \Gamma_S \circ +$ is given on $\prod P_s$ by "coordinatewise multiplication." The properties of Γ_S which interest us are

Lemma 4.3:

1) *Let* $\sigma : \prod P_s \to \coprod P_s$ *be a section of* γ, $\eta : 1 \xrightarrow{\sim} \sigma \circ \gamma$, *then*

$$\Gamma_S(\eta) : \Gamma_S \xrightarrow{\sim} \Gamma_S \circ \sigma \circ \gamma$$

(trivial).

2) $$\Gamma_S = \prod \Gamma_s \circ \gamma$$

(definition).

3) *If* Q *is a permutative category*, $\wedge : Q \to \underline{\underline{C}}$ *a permutative functor*, $\alpha_s : P_s \to Q$ *functors of permutative categories, and*

$\lambda_s : \Gamma_s \to \wedge \circ \alpha_s$ natural transformations of permutative functors, then there exists a unique natural transformation of permutative functors

$$\coprod_S \lambda_s : \Gamma_S \longrightarrow \wedge \circ \coprod_S \alpha_s$$

where $\coprod_S \alpha_s : \coprod_S P_s \longrightarrow Q$ is the natural extension of the α_s .

4) As a consequence of 3), Γ_S is functorial in S .

The proof of this lemma is messy but straightforward.

We may now complete the definition of our chain functors. Let $\Gamma : \Sigma \to \underline{\underline{C}}$ be a Σ-structure. Denote also by Γ the induced permutative functor $\tilde{\Gamma} : \tilde{\Sigma} \to \underline{\underline{C}}$. For a finite pointed set S , let

$$P_s = \begin{cases} \tilde{\Sigma} , & s \neq * \\ * , & s = * \end{cases}$$

Then we have

$$\Gamma_S : \coprod_S P_s \longrightarrow \underline{\underline{C}}$$

Let $B : \underline{\underline{C}} \longrightarrow$ (pointed spaces) be a functor such that

1) $B(X \times Y) \xrightarrow{\sim} B(X) \times B(Y)$

2) $B* = *$.

To define $\Phi(S)$ we take a homotopy limit of $B \circ \Gamma_S$, [T2].

$$\Phi_{\Gamma,B}(S) = \underset{\overrightarrow{\coprod_S P_s}}{\text{holim}}\, B\Gamma_S$$

Then an easy consequence of 4.3 is

Theorem 4.4. $\Phi_{\Gamma,B}$ is a chain functor.

We can easily verify that for $\Gamma = BG\ell(n,F)$ or $\Gamma = BG\ell(n,\mathbb{C})$ and $B = \mathrm{id}$, $\Phi_{\Gamma,B}$ is homotopy equivalent to the chain functors given by the constructions of Segal and Anderson.

To get the theory $K_{et}\overline{\mathbb{F}}_q(\ \ )$ desired for §1, let $\underline{\underline{C}}$ = (group schemes / $\overline{\mathbb{F}}_q$), $\Gamma\underline{n} = G\ell(n,\overline{\mathbb{F}}_q)$. BG_{et}, the étale classifying homotopy type of $G \in \underline{\underline{C}}$, is a pro-space. To get a suitable functor B we need the homotopy inverse limit of a suitable completion of $B_{et}(\ \ )$. Specifically

$$B = (\underleftarrow{\mathrm{holim}}\, B_{et})[\tfrac{1}{p}], \quad p = \mathrm{char}\, \mathbb{F}_q .$$

Then B preserves products up to homotopy.

§5. Further applications.

Quillen has suggested to us that our techniques could be used to extend the Quillen-Friedlander proof of the Adams conjecture to the following stable version.

Conjecture 6.1. The diagram

$$\begin{array}{ccc} \underline{\underline{KC}}[\tfrac{1}{p}] & \xrightarrow{\ \psi^p\ } & \underline{\underline{KC}}[\tfrac{1}{p}] \\ & {}_{J}\searrow \quad \swarrow_{J} & \\ & \underline{\underline{BG}}[\tfrac{1}{p}] & \end{array}$$

homotopy commutes.

The details indeed appear to go through, but are sufficiently complicated that we do not wish to claim this conjecture as a theorem in this note. The essential point seems to be that all the constructions in Quillen's proof be functorial.

A second potential application is the construction of new cohomology theories from algebraic groups over fields that are not algebraically closed. Theories of that kind would contain additional information coming from Galois group actions.

REFERENCES

Topological:

T1. D. W. Anderson, "Chain functors and homology theories," Symposium on Algebraic Topology, Battelle, 1971., pp. 1-12, Lecture Notes in Mathematics Vol. 249, Springer-Verlag, New York.

T2. A. K. Bousfield and D. M. Kan, Homotopy limits, completions and localizations, Lecture Notes in Mathematics, Vol. 304, Springer-Verlag, New York, 1972.

T3. D. G. Quillen, "Some remarks on étale homotopy theory and a conjecture of Adams," Topology 7 (1968), 111-116.

T4. ____________, "On the cohomology and K-theory of the general linear groups over a finite field," Annals of Math., 96(1972), 552-586.

T5. G. Segal, "Homotopy everything H-spaces," to appear.

T6. J. Tørnehave, "Delooping the Quillen map," Thesis, M.I.T., 1971.

Geometric:

G1. M. Artin, Grothendieck Topologies, Harvard lecture notes, 1962.

G2. ____________, A. Grothendieck, and J. L. Verdier, Seminaire de Géométrie Algébrique du Bois Marie 1963/64, SGA4, Lecture Notes in Mathematics nos. 269, 270, 305, Springer-Verlag, New York. 1971.

G3. A. Grothendieck, "Classes de Chern et représentations linéaires des groups discrets," Dix Exposés sur la Cohomologie des Schémas, pp. 215-305, North Holland, Amsterdam, 1968.

G4. L. Illusie, "Travaux de Quillen sur la cohomologie des groupes," Semenaire Bourbaki, 405, Lecture notes in mathematics Vol. 317, Springer-Verlag, New York.

G5. I. G. MacDonald, Algebraic Geometry, Benjamin 1968.

G6. D. Mumford, Introduction to Algebraic geometry (preliminary version), Harvard lecture notes.

RELATIONS IN REGULAR CATEGORIES

Jeanne Meisen

Case Western Reserve University

1. Introduction

Canonical factorization of morphisms is a common and important concept in category theory. It received early attention with the appearance of abelian categories. Later, G. M. Kelly introduced a general notion of $(\underline{E},\underline{M})$ factorizations [6,10], defined below, and in his unpublished notes, he discussed several interesting examples of unusual factorizations where $\underline{M}$ need not be a class of monomorphisms. One such example is in the category of abelian groups, where $\underline{M} = \{f/$ coker f is torsion free and ker f is divisible and torsion free$\}$ and $\underline{E} = \{f/$ coker f and ker f are torsion$\}$.

An $(\underline{E},\underline{M})$ factorization of morphisms in a category $\mathcal{A}$ is defined as a pair of classes of morphisms in $\mathcal{A}$ satisfying the following axioms:

(1) Every isomorphism is in both $\underline{E}$ and $\underline{M}$.

(2) $\underline{E}$ and $\underline{M}$ are closed under composition.

(3) For any commutative square,

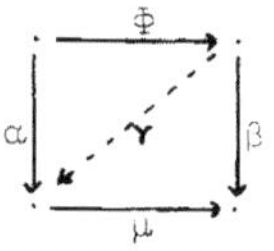

in $\mathcal{A}$

with $\Phi \in \underline{E}$ and $\mu \in \underline{M}$, there exists a unique γ such that $\gamma\Phi = \alpha$ and $\mu\gamma = \beta$.

(4) For every morphism α in $\mathcal{A}$, there exist $\alpha_m \in \underline{M}$ and $\alpha_e \in \underline{E}$ such that $\alpha = \alpha_m \alpha_e$.

It is readily seen that the $(\underline{E},\underline{M})$ factorization (4) is essentially unique.

Let P be a family of primes and let $\mathcal{A}$ be any category in which we have a P-localization theory, thus $\mathcal{A}$ might be the category of nilpotent groups or the category of nilpotent spaces [12]. Abstract from $\mathcal{A}$ the subcategory $\mathcal{A}(P)$ whose objects are those of $\mathcal{A}$ and whose morphisms are the identities of $\mathcal{A}$ together with the morphisms of $\mathcal{A}$ whose codomains are P-local. There is then a canonical $(\underline{E},\underline{M})$ factorization on $\mathcal{A}(P)$, where $\underline{M}$ consists of identities and morphisms between P-local objects and $\underline{E}$ consists of identities and P-localizing morphisms $e: A \to A_P$. It is plain that conditions (1) and (2) for an $(\underline{E},\underline{M})$ factorization are satisfied. Moreover, conditions (3) and (4) follow from the universal property of P-localization which guarantees, for each $\varphi: A \to B$ with B P-local, a unique morphism $\psi: A_P \to B$ with $\psi e = \varphi$.

Indeed, there is a converse to the above. For given the classes $(\underline{E},\underline{M})$ above, defined on the subcategory $\mathcal{A}(P)$ of an arbitrary category $\mathcal{A}$ in which a notion of P-local objects has been introduced, then we have a P-localization theory on $\mathcal{A}$ if and only if $(\underline{E},\underline{M})$ satisfies (3) and (4). Thus we may expect to find a close connection between localization theory and the theory of $(\underline{E},\underline{M})$ factorizations; and it is hoped to devote a later paper to a study of this connection.

In [10] we began to study relations in a category with finite products and an $(\underline{E},\underline{M})$ factorization. A relation is defined to be $A \xleftarrow{\alpha} R \xrightarrow{\beta} B$ with $\{\alpha,\beta\} \in \underline{M}$. In this paper we shall specialize to a regular category [1,3] which is a finitely complete category with ($\underline{E}$ = regular epis, $\underline{M}$ = monos) factorization, such that the pullback preserves $\underline{E}$ - morphisms. Proofs will be omitted and readers are referred to [11] for details.

In the next section we give some preliminary definitions. Relations form the morphisms of a bicategory of relations, Rel $\mathcal{A}$. We can, up to isomorphism of morphisms, recapture the original category $\mathcal{A}$ from Rel $\mathcal{A}$. In Section 3 we show that the passage from $\mathcal{A}$ to Rel$\mathcal{A}$ does not render invertible any morphism not already invertible. Thus, for any relation R, R is invertible if and only if $R \circ \bar{R} \sim I$, $\bar{R} \circ R \sim I$ where $\bar{R}$ is the <u>converse</u> relation to R. These considerations lead naturally to the study of difunctional relations. A relation R is called <u>difunctional</u> (or <u>von Neumann regular</u>) if $R \circ \bar{R} \circ R \sim R$, and in Section 4 we show that all pullback relations are difunctional. It is known that in abelian categories all relations are pullbacks and difunctional [4]: in the category of groups, all relations are difunctional [8] but not necessarily pullbacks; and in the category of M-sets, not every relation is difunctional but every difunctional relation is a pullback relation. We therefore study the condition for all difunctional relations to be pullbacks in Section 5. In the last section, we generalize some known results [2] relating difunctional and equivalence relations in algebraic categories, to exact categories [1].

2. The bicategory Rel $\mathcal{A}$

Let $\mathcal{A}$ be a regular category with pushouts. A triple (R,α,β) with objects R, A, B, and morphisms $\alpha\colon R \longrightarrow A$, $\beta\colon R \longrightarrow B$ in $\mathcal{A}$ is a relation from B to A when $\{\alpha,\beta\} : R \longrightarrow A \times B$ is mono. We abbreviate (R,α,β) by R whenever there is no ambiguity.

Composition of relations R and S, with (S,γ,δ) from C to B, is defined by by the rule $(R,\alpha,\beta) \circ (S,\gamma,\delta) = (R \circ S, (\alpha\mu)', (\delta\nu)')$ where $R \xleftarrow{\mu} P \xrightarrow{\nu} S$ is the pullback of $R \xrightarrow{\beta} B \xleftarrow{\gamma} S$ and $\{\alpha\mu,\delta\nu\} = \{(\alpha\mu)', (\delta\nu)'\}\,\epsilon$ is the factorization of $P \longrightarrow A \times C$. This composition is associative in regular categories but not in general [3,7].

Let (R,α,β) and (S,γ,δ) be relations from B to A. A map from (R,α,β) to (S,γ,δ) is a commutative diagram in $\mathcal{A}$;

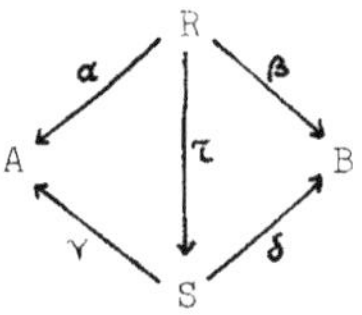

such that $\gamma\tau = \alpha$ and $\delta\tau = \beta$. We note that τ is uniquely determined and is mono. The category of relations from B to A with maps between relations as morphisms, $\mathrm{Rel}(A,B)$, is therefore a preordered set; we denote the preorder by $\subseteq$.

(R,α,β) and (S,γ,δ) are said to be isomorphic if and only if τ is an isomorphism. Then relations as defined above form the morphisms of a bicategory, Rel $\mathcal{A}$ [10]. There is an embedding functor $F\colon \mathcal{A} \to \mathrm{Rel}\mathcal{A}$ with $F(A) = A$ and

$F(\alpha) = (R,\alpha,1_R)$ where $\alpha: R \longrightarrow A$ in $\mathcal{A}$.

3. Converse relations and adjoint relations

To any relation (R,α,β) from B to A, there is a converse relation (R,β,α), denoted by $\bar{R}$, from A to B. Obviously we have (i) $\bar{\bar{R}} \sim R$ and (ii) $\overline{R \circ S} \sim \bar{S} \circ \bar{R}$ where (S,γ,δ) is a relation from C to B; here $\sim$ denotes isomorphism.

Proposition 3.1. (i) $R \circ \bar{R} \subseteq I_A \Leftrightarrow \beta$ mono,
(ii) $R \circ \bar{R} \supseteq I_A \Leftrightarrow \alpha$ regular epi.

Corollary 3.2. $R \circ \bar{R} \sim I_A \Leftrightarrow \beta$ mono and α regular epi.

In Rel $\mathcal{A}$, a relation $R: B \longrightarrow A$ is said to have a right adjoint $S: A \longrightarrow B$ if there exist $\eta: I_B \longrightarrow S \circ R$ and $\tau: R \circ S \longrightarrow I_A$ such that
$R \xrightarrow{R\eta} R \circ S \circ R \xrightarrow{\tau R} R = 1$, $S \xrightarrow{\eta S} S \circ R \circ S \xrightarrow{S\tau} S = 1$.

The above equations follow automatically from the existence of η and τ , since Rel(A,B) forms a preordered set.

Proposition 3.3. If $(R,\alpha,\beta): B \to A$ has a right adjoint S then $S = \bar{R}$. Moreover β is then an iso.

We conclude:

Theorem 3.4. The following are equivalent:

(i) $R \circ \bar{R} \subseteq I_A$, $\bar{R} \circ R \supseteq I_B$,
(ii) β is an iso ,
(iii) $(R,\alpha,\beta) \sim (R,\rho,1)$,
(iv) R has a right adjoint.

Thus we can, up to isomorphism of morphisms, recapture the original category $\mathcal{A}$ from Rel$\mathcal{A}$. Furthermore, let a relation (R,α,β) be an equivalence (i.e. there exists a relation S such that $R \circ S \sim I$ and $S \circ R \sim I$). Then R has a two sided adjoint $S \sim \bar{R}$ and α,β are iso. Hence $(R,\alpha,\beta) \sim (B,\rho,1)$ with ρ iso. Thus:

Corollary 3.5. The equivalence morphisms of Rel $\mathcal{A}$ are just the isomorphisms of $\mathcal{A}$.

4. Difunctional relations

We call a relation difunctional if $R \circ \bar{R} \circ R \sim R$ and define a pullback relation to be a relation which is a pullback of some pair of morphisms. We note that a pullback is always a pullback of its own pushout and is always a relation.

Theorem 4.1. Every pullback relation is difunctional.

We will discuss the converse of Theorem 4.1 in the next section. As a consequence of Corollary 3.2 and Theorem 4.1, we can characterize the monomorphisms and epimorphisms among the pullback relations in Rel $\mathcal{A}$.

Theorem 4.2. Let (R,α,β) be a pullback relation. Then the following are equivalent:

(i) R is mono
(ii) $\bar{R}$ is epi
(iii) $\bar{R} \circ R \sim I$
(iv) α is mono and β is regular epi.

Corollary 4.3. A pullback relation is an equivalence if and only if it is mono and epi.

Theorem 4.4. Every pullback relation R has a canonical factorization $R \sim M \circ E$ with M mono and E epi in Rel $\mathcal{A}$.

5. E-relations

We call a relation (R,α,β) with α,β regular epi an E-relation. Associated with each relation (R,α,β), there is a canonical E-relation (R,α_e,β_e), denoted by R'. If R is a pullback relation, so is R'. The converse is not always true.

An exact category is a regular category in which every equivalence relation is a kernel pair. In the following theorem we give a condition for the converse of Theorem 4.1.

Theorem 5.1. Let $\mathcal{A}$ be an exact category with pushouts. Suppose that $R = (R,\alpha,\beta)$ is a relation and let R' be its E-relation. Then the following are equivalent:

(i) R' is a pullback relation,
(ii) R is difunctional,

(iii) R' is difunctional,

(iv) $R \sim M \circ E$ where $M = (M, \alpha_m, \varphi)$ and $E = (E, \Psi, \beta_m)$ with φ, Ψ regular epi,

(v) $R' \sim M' \circ E'$

Theorem 5.2. Let $\mathcal{A}$ be an exact category with pushouts. Suppose every pair $\leftarrowtail \cdot \longrightarrow$ with $\leftarrowtail$ mono in $\mathcal{A}$ is a pullback. Then R is a pullback if it is difunctional.

In the category of groups, every relation is difunctional. In the following example we show that in this category, not every (difunctional) relation is a pullback.

Example: Take a nonabelian simple group G and a non-trivial cyclic subgroup A of G. Then $G \leftarrowtail A \longrightarrow \{1\}$ is a relation but its pushout diagram

$$\begin{array}{ccc} G & \longrightarrow & P \\ \uparrow & & \uparrow \\ A & \longrightarrow & \{1\} \end{array}$$

has $P \sim G$ and is not a pullback.

6. Equivalence relations

We generalize some results known for algebraic categories to exact categories

Theorem 6.1. In an exact category with pushouts, the following are equivalent:

(i) Every relation is difunctional.

(ii) Every reflexive relation is an equivalence relation.

(iii) The composite of two equivalence relations is an equivalence relation.

(iv) Equivalence relations on the same object commute.

REFERENCES

1. Barr, M. Exact categories, Lecture Notes in Math. 236, Springer-Verlag, (1971) pp. 1-120.

2. Findlay, G. D. Reflexive Homomorphic Relations, Can. Math. Bull. 3 No. 2 (1960) 131-132

3. Grillet, P. A. Regular categories, Lecture Notes in Math. 236, Springer-Verlag (1971) pp. 121-222

4. Hilton, P. J. Correspondences and exact squares, Proc. of the Conferences on Categorical Algebra, La Jolla, Calif. (1965), pp. 254-271.

5. Hilton, P., Mislin, G., and Roitberg, J. Localization of nilpotent groups and spaces, North Holland (1974)

6. Freyd, P. J., and Kelly, G. M. Categories of continuous functors I, J. of Pure and Applied Algebra, 2, (1972), pp. 169-191.

7. Klein, A. Relations in categories, Ill. J. Math. 14 (1970) pp. 536-550.

8. Lambek, J. Goursat's theorem and the Zassenhaus lemma, Can. J. Math., 10 (1957), pp. 45-56.

9. MacLane, S. An algebra of additive relations, Proc. Nat. Acad. Sci., 47 (1961), pp. 1043-1051.

10. Meisen, J. On bicategories of relations and pullback spans, Communications in Algebra 1 (5), (1974) pp. 377-401.

11. Meisen, J. On some properties of relations in regular categories (to appear)

12. Puppe, D. Korrespondenzen in Abelschen Kategorien, Math. Ann. 148 (1962) pp. 1-30.

Nilpotent groups with finite commutator subgroups

Guido Mislin

Eidgenössische Technische Hochschule, Zürich

The object of this note is to investigate non-cancellation phenomena in the category of finitely generated nilpotent groups. Our main tool is the theory of localization of nilpotent groups, a technique which has its roots in the papers [6], [5] of Malcev and Lazard respectively. Localization of nilpotent groups has recently been discribed extensively in several places [1], [2], [4], [11], [12]; the basic results relevant to the applications presented here may be found in Hiltons paper [2]. We will denote by N_p the p-localization of a nilpotent group N, and we will write N_o for its rationalization (Malcev completion).

It has been observed by Milnor [7] that there exist non-isomorphic finitely generated nilpotent groups, which have isomorphic p-localizations for all primes p. This prompts the following basic definition.

Definition For a finitely generated nilpotent group N, the genus G(N) consists in the family of all isomorphism classes of finitely generated nilpotent groups M with M_p isomorphic to N_p for all primes p.

It follows from results of Pickel [10], that G(N) is always a finite set; notice that his notion of "genus" differs slightly from ours.

For a finite homotopy associative H-complex X and a finite complex Y; the set of homotopy classes of maps $Y \to X$ forms a finitely generated nilpotent group with finite commutator subgroup. This fact, in conjunction with the results of [3], [8], [9], [13] and [14], may be considered as motivation for the following study.

The main results are the following.

Theorem 1 Let M and N be finitely generated nilpotent groups, and let A be a finitely generated abelian group. Then $G(M\times A) = G(N\times A)$ implies that $G(M) = G(N)$.

Theorem 2 Let M and N be finitely generated nilpotent groups with finite commutator subgroups. Then the following are equivalent.

(i) $G(M) = G(N)$

(ii) $M \times \mathbb{Z}^{h(M)} \cong N \times \mathbb{Z}^{h(N)}$; h the Hirsch number

(iii) There exists a finitely generated abelian group A such that $M \times A \cong N \times A$.

For N as in Theorem 2, we will give a precise description of the genus set $G(N)$, as a subset of a certain space of double cosets (Theorem 4), and we will deduce from this description a finite upper bound for the cardinality of $G(N)$. Moreover, we will show that for M in $G(N)$, some (cartesian) power of M is isomorphic to the same power of N (Theorem 6).

The next theorem will provide non trivial examples of nilpotent groups fulfilling the hypothesis of Theorem 2.

Theorem 3 Let p denote an odd prime and let

$$\{x_i, y_i \mid 1 \leq i \leq (p-1)/2\} \subset H^1(\mathbb{Z}^{p-1}; \mathbb{Z}/p)$$

be the reduction mod p of a basis of $H^1(\mathbb{Z}^{p-1}; \mathbb{Z}) \cong \mathbb{Z}^{p-1}$. Consider the central extension

$$E(p) : \mathbb{Z}/p \rightarrowtail N(p) \twoheadrightarrow \mathbb{Z}^{p-1}$$

with $[E(p)] = \Sigma x_i y_i \in H^2(\mathbb{Z}^{p-1}; \mathbb{Z}/p)$. Then

$$|G(N(p))| = \frac{p-1}{2} .$$

Notice that the isomorphism class of $N(p)$ is clearly independent of the basis chosen in $H^1(\mathbb{Z}^{p-1}; \mathbb{Z})$, and the commutator subgroup of $N(p)$ is cyclic of order p.

Proof of Theorem 1 It is of course sufficient to prove the theorem in case that A is cyclic. Hence we can assume that for all primes p, the ring of endomorphisms $\mathrm{End}(A_p)$ is a local ring. For a homomorphism $g : U \times V \to S \times T$ we write $g(U,T)$ for $pr_T \circ g \circ in_U$. Let $f : M_p \times A_p \to N_p \times A_p$ denote an isomorphism. We will distinguish two cases

1) $f(A_p,A_p)$ is an isomorphism. Then it is easily checked that $f^{-1}(N_p,M_p)$ is an isomorphism, with inverse $f(M_p,N_p) - f(A_p,N_p)\, f(A_p,A_p)^{-1} f(M_p,A_p)$; if $\varphi,\psi : G \to H$ are maps between groups, we denote by $\varphi \pm \psi$ the map $g \mapsto \varphi(g)\psi(g^{\pm 1})$.

2) $f(A_p,A_p)$ is not an isomorphism. Then $f^{-1}(A_p,A_p)f(A_p,A_p)$ is not an isomorphism either, since A_p is a finitely generated $\mathbb{Z}_p$-module. But

$$f^{-1}(A_p,A_p)f(A_p,A_p) + f^{-1}(N_p,A_p)f(A_p,N_p) = \mathrm{Id}$$

Since $\mathrm{End}(A_p)$ is a local ring, it follows that $f^{-1}(N_p,A_p)f(A_p,N_p)$ must be an isomorphism (i.e. a unit in $\mathrm{End}(A_p)$). Hence $f(A_p,N_p) : A_p \to N_p$ has a left inverse. Since A_p lies in the center of $M_p \times A_p$ and since

$$M_p \times A_p \overset{f}{\rightarrowtail} N_p \times A_p \overset{pr}{\twoheadrightarrow} N_p$$

maps the center of $M_p \times A_p$ into the center of N_p, it follows that $f(A_p,N_p)$ maps A_p onto a central retract of N_p. If K denotes the kernel of the retraction map $N_p \twoheadrightarrow A_p$, then $N_p = K \times A_p$. Hence there is an automorphism λ of $N_p \times A_p = K \times A_p \times A_p$ such that $\lambda f : M_p \times A_p \rightarrowtail\!\!\!\!\twoheadrightarrow N_p \times A_p$ has the property that $(\lambda f)(A_p,A_p)$ is an isomorphism, and we are back in case 1.

An immediate consequence of Theorem 1 is

Corollary 1. Let M and N be finitely generated nilpotent groups and let A be a finitely generated abelian group. Then $M \times A \cong N \times A$ implies that $G(M) = G(N)$.

As usual, we denote by $[N,N]$ the commutator subgroup of N. Notice that the converse of the conclusion of Corollary 1 cannot hold in general, since $M \times A \cong N \times A$ implies that $[M,M] \cong [N,N]$. However, under more restrictive conditions on the groups involved, one can prove a converse (cf. Theorem 2). For this, we need first a couple of preliminary lemmas.

Denote the center of a group G by $Z(G)$.

Lemma 1 Let N be a finitely generated nilpotent group with finite commutator subgroup. Then $N/Z(N)$ is a finite group and the Hirsch number $h(N)$ of N is given by

$$h(N) = \text{rank}\ (Z(N)) = \dim_{\mathbb{Q}} N_o$$

Proof. Consider the commutative diagram with exact rows

$$\begin{array}{ccccc} Z(N) & \rightarrowtail & N & \twoheadrightarrow & N/Z(N) \\ \downarrow & & \downarrow & & \downarrow \\ Z(N)_o & \rightarrowtail & N_o & \twoheadrightarrow & (N/Z(N))_o \end{array}$$

The vertical arrows are the rationalization maps. Since N is finitely generated, we can identify $Z(N_o)$ with $Z(N)_o$ (cf. [2]).

N_o is abelian because the (finite) commutator subgroup $[N,N]$ lies in the kernel of the rationalization map $N \to N_o$. Hence $Z(N_o) = N_o$ and $(N/Z(N))_o = \{1\}$. It follows that $N/Z(N)$ is finite. Of course this implies that the Hirsch number $h(N) = \text{rank}\ (Z(N)) = \dim_{\mathbb{Q}} N_o$.

A further property of finitely generated nilpotent groups with finite commutator subgroups is the following.

<u>Lemma 2</u> Let N be a finitely generated nilpotent group with finite commutator subgroup. Then $M \in G(N)$ implies that $[M,M] \cong [N,N]$.

<u>Proof.</u> Recall that for any nilpotent group N the canonical map $[N,N] \to [N_p,N_p]$ p-localizes. Hence $[N_p,N_p] \cong [N,N]_p$. Further, since $[N,N]$ is finite

$$[N,N] \cong \prod_p [N,N]_p$$

For $M \in G(N)$ one has $[M_p,M_p] \cong [N_p,N_p]$. Hence, it follows immediately that $[M,M]$ is finite and $[M,M] \cong \Pi[M_p,M_p] \cong \Pi[N_p,N_p] \cong$ $\cong [N,N]$.

Denote by $T(N)$ the torsion subgroup of a nilpotent group N.

<u>Definition</u> If N is a nilpotent group with $T(Z(N))$ a finite group, then

$$FZ(N) = \{x \in Z(N) \mid \exists\ y \in Z(N) \text{ with } y^n = x,\ n = |T(Z(N))|\}$$

Notice that $FZ(N)$ is a torsionfree abelian group; if N is finitely generated with $[N,N]$ finite, then $FZ(N)$ is a free abelian group of rank $h(N)$. If $f : N \to M$ is an isomorphism of nilpotent groups and $TZ(N)$ is finite, then f induces an isomorphism $FZ(N) \to FZ(M)$; in particular, $FZ(N)$ is a characteristic subgroup of N. Further, if N is finitely generated the canonical map $N \to N_p$ maps $FZ(N)$ into $FZ(N_p)$, since it maps $Z(N)$ into $Z(N_p)$ and since $|TZ(N_p)|$ divides $|TZ(N)|$; hence $FZ(N)_p$ may be identified with a subgroup of $FZ(N_p)$. Moreover one has

<u>Lemma 3</u> If N is finitely generated nilpotent, then the canonical map

$$FZ(N)_p \to FZ(N_p)$$

maps $FZ(N)_p$ isomorphically onto the characteristic subgroup

$$X = \{x \in Z(N_p) \mid \exists\, y \in Z(N_p) \text{ with } y^n = x,\ n = |TZ(N)|\} \subset N_p$$

<u>Proof</u>. One has a commutative diagram with exact rows

$$\begin{array}{ccccc}
TZ(N) & \rightarrowtail & Z(N) & \xrightarrow{n} & FZ(N) \\
\downarrow{\scriptstyle\alpha} & & \downarrow{\scriptstyle\beta} & & \vdots\,{\scriptstyle\gamma} \\
TZ(N_p) & \rightarrowtail & Z(N_p) & \xrightarrow{n} & X
\end{array}$$

Since N is finitely generated, α and β p-localize (cf. [2]). Hence γ p-localizes (cf. [2]).

Denote the quotient group $N/FZ(N)$ by $Q(N)$.

<u>Lemma 4</u> Let N be a finitely generated nilpotent group with $[N,N]$ finite. Then $M \in G(N)$ implies that $FZ(M) \cong FZ(N)$ and $Q(M) \cong Q(N)$.

<u>Proof</u>. For $M \in G(N)$ one has $M_o \cong N_o$ and hence, since $[N,N]$ and $[M,M]$ are finite, $h(M) = h(N) = h$ by Lemma 1. So $FZ(M) \cong FZ(N) \cong Z^h$. Let $f : M_p \to N_p$ be an isomorphism. Since $M \in G(N)$ implies that $Z(M) \in G(Z(N))$ and hence $Z(M) \cong Z(N)$, we infer by virtue of Lemma 3, that f maps $FZ(M)_p$ isomorphically onto $FZ(N)_p$. Hence there is a commutative diagram with exact rows

$$\begin{array}{ccccc}
FZ(M)_p & \rightarrowtail & M_p & \twoheadrightarrow & Q(M)_p \\
\wr\downarrow{\scriptstyle\bar{f}} & & \wr\downarrow{\scriptstyle f} & & \vdots \\
FZ(N)_p & \rightarrowtail & N_p & \twoheadrightarrow & Q(N)_p
\end{array}$$

It follows that $Q(M)_p \cong Q(N)_p$ for all primes p. Since, by Lemma 1, $N/Z(N)$ is finite, and since $FZ(N)$ has finite index in $Z(N)$, it follows that $Q(N) = N/FZ(N)$ is a finite group; similarly for $Q(M)$. Hence $Q(N) \cong \Pi\, Q(N)_p \cong \Pi\, Q(M)_p \cong Q(M)$.

We will now use Lemma 4 to prove Theorem 2.

Proof of Theorem 2. Trivially (ii) implies (iii) and, by Corollary 1, (iii) implies (i). It remains to show that (i) implies (ii). Under the conditions stated, $G(M) = G(N)$ implies, using Lemma 4 , that both M and N are central extensions of a finite group Q by Z^h, where

$$Q \cong Q(M) \cong Q(N) \quad \text{and} \quad h = h(M) = h(N) .$$

Let

$$E \;:\; Z^h \rightarrowtail M \twoheadrightarrow Q$$

$$E'' \;:\; Z^h \rightarrowtail N \twoheadrightarrow Q$$

denote two such extensions, in which the image of Z^h is $FZ(M)$ and $FZ(N)$ respectively; $[E], [E''] \in H^2(Q; Z^h)$. Since $M_p \cong N_p$ for all primes p, and since every isomorphism $M_p \to N_p$ maps $FZ(M)_p$ into $FZ(N)_p$, there exist isomorphisms $\kappa(p): Z_p^h \to Z_p^h$ and $\lambda(p): Q_p \to Q_p$ such that

$$\kappa(p)_* \,[E]_p = \lambda(p)^* \,[E'']_p \quad \text{in} \quad H^2(Q_p; Z_p^h)$$

Let $\lambda = \Pi\lambda(p) : Q \to Q$ be the global automorphism with $\lambda_p = \lambda(p)$ for all primes p. Then $\lambda^*[E''] = [E']$ where

$$E' \;:\; Z^h \rightarrowtail \bar{N} \twoheadrightarrow Q$$

is an extension, representing a group $\bar{N}$ isomorphic to N. Denote the projections $M \to Q$ in E by pr_M, and $\bar{N} \to Q$ in E' by $pr_{\bar{N}}$. Notice that

$$\kappa(p)_*[E]_p = [E']_p \quad \text{for all primes } p.$$

Consider the diagram

$$(*)\qquad\begin{array}{ccccc} & & & & Z^h \\ & & & & \downarrow \\ & & W & \dashrightarrow & \bar{N} \\ & & \vdots & & \downarrow{\scriptstyle pr_{\bar{N}}} \\ Z^h & \longrightarrow & M & \xrightarrow[pr_M]{} & Q \end{array}$$

in which W denotes the pullback of pr_M and $pr_{\bar{N}}$. Since $E : Z^h \to M \to Q$ gives rise to an exact sequence [11]

$$Hom(Z^h, Z^h) \xrightarrow{\delta} H^2(Q; Z^h) \xrightarrow{pr_M^*} H^2(M; Z^h)$$

with $\delta(Id) = [E]$, it follows that $pr_M^*[E] = 0$. Similarly $pr_{\bar{N}}^*[E'] = 0$. Our aim is to prove that, in addition, $pr_{\bar{N}}^*[E] = 0$ and $pr_M^*[E'] = 0$. It suffices, of course, to check those equalities at each individual prime p : $(pr_M^*[E'])_p = pr_{M_p}^*[E']_p = pr_{M_p}^*\kappa(p)_*[E]_p = \kappa(p)_* pr_{M_p}^*[E]_p = \kappa(p)_*(pr_M^*[E])_p = 0$, since already $pr_M^*[E] = 0$. Similarly, $pr_{\bar{N}}^*[E] = 0$. It follows in the diagram (*) that $W \cong M \times Z^h$ and simultaneously $W \cong \bar{N} \times Z^h$. Hence the result.

We will now discuss in more detail the genus set $G(N)$ for N a finitely generated nilpotent group with finite commutator subgroup. We have already seen that $M \in G(N)$ may be represented by a central extension $E : Z^h \xrightarrow{f} M \to Q$ with $f(Z^h) = FZ(M)$ and $Q \cong Q(M)$. The element $[E] \in H^2(Q;Z^h)$ is, of course, not uniquely determined by the isomorphism class of M; if $\bar{E} : Z^h \xrightarrow{\bar{f}} \bar{M} \to Q$ is another central extension with $\bar{f}(Z^h) = FZ(\bar{M})$ and $M \cong \bar{M}$, then $[\bar{E}]$ and $[E]$ lie in the same double coset of the right $Aut\ Q$ - left $Aut\ Z^h$ action on $H^2(Q;Z^h)$ and, vice versa, every $[E']$, $E' : Z^h \xrightarrow{g} L \to Q$, which lies in the same double coset as $[E]$, will give rise to a commutative diagram

$$\begin{array}{ccccccc} E' : & Z^h & \xrightarrow{g} & L & \longrightarrow & Q \\ & \cong\downarrow & & \vdots & & \downarrow\cong \\ E : & Z^h & \xrightarrow{f} & M & \longrightarrow & Q \end{array}$$

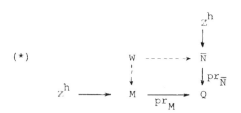

from which it follows that $L \cong M$ and that $g(\mathbb{Z}^h) = FZ(L)$. Hence, there must exist a bijection of a subset of the space of double cosets of $H^2(Q;\mathbb{Z}^h)$ onto $G(N)$. Since Q is finite, one has $H^2(Q;\mathbb{Z}^h) \cong H^1(Q;\mathbb{R}^h/\mathbb{Z}^h) \cong \mathrm{char}(Q)^h$, where $\mathrm{char}(Q)$ is the group of characters $Q \to S^1$. Hence we see that $H^2(Q;\mathbb{Z}^h)$ has exponent t where

$$t = \exp(Q_{ab}) \quad ; \quad Q_{ab} = Q/[Q,Q]$$

Further, since Q is finite nilpotent, $p/|Q|$ iff p/t. Thus, $\mathbb{Z} \to \mathbb{Z}_t = \bigcap_{p/t} \mathbb{Z}_p$ induces an isomorphism

$$H^2(Q;\mathbb{Z}^h) \overset{\cong}{\to} H^2(Q;\mathbb{Z}_t^h)$$

It follows that $H^2(Q;\mathbb{Z}^h)$ is equipped with a left $G = GL(h,\mathbb{Z})$ action, a left $G(t) = GL(h,\mathbb{Z}_t)$ action, and a right $\tilde{Q} = \mathrm{Aut}\, Q$ action; the injection $GL(h,\mathbb{Z}) \to GL(h,\mathbb{Z}_t)$ gives rise to a surjection of double coset spaces

$$\sigma : G\backslash H^2(Q;\mathbb{Z}^h)/\tilde{Q} \twoheadrightarrow G(t)\backslash H^2(Q;\mathbb{Z}^h)/\tilde{Q}$$

$$G[E]\,\tilde{Q} \to G(t)[E]\tilde{Q}$$

<u>Theorem 4</u> Let N be a finitely generated nilpotent group with finite commutator subgroup, and let

$$E : \mathbb{Z}^h \overset{f}{\to} N \to Q$$

be a central extension with $f(\mathbb{Z}^h) = FZ(N)$ and $Q \cong Q(N)$. Then there is a natural bijection

$$\theta : \sigma^{-1}(G(t)[E]\tilde{Q}) \to G(N)$$

<u>Proof</u>. Let $E' : \mathbb{Z}^h \overset{f'}{\to} M \to Q$ represent an element of $\sigma^{-1}(G(t)[E]\tilde{Q})$. We would like to define θ by $\theta[E'] = M$. In order that this makes sense, we first have to check that $M \in G(N)$. Denote by M_t the localization M_P for $P = \{p|p/t\}$; similarly for N_t. Since $G(t)[E']\tilde{Q} = G(t)[E]\tilde{Q}$, there exist $\alpha \in G(t)$ and $\beta \in \tilde{Q}$ such that $\alpha_*[E']_t = \beta^*[E]_t$ in $H^2(Q;\mathbb{Z}_t^h)$, and hence there exists an isomorphism

$\gamma : M_t \to N_t$ making the following diagram commutative

$$\begin{array}{ccccccc} E'_t & : & Z_t^h & \xrightarrow{f'_t} & M_t & \longrightarrow & Q \\ & & \alpha\downarrow & & \gamma\downarrow & & \downarrow\beta \\ E_t & : & Z_t^h & \longrightarrow & N_t & \longrightarrow & Q \end{array}$$

It follows that $M_p \cong N_p$ for all primes p which divide t. If q is a prime which does not divide t, one has clearly $M_q \cong Z_q^h \cong N_q$. Hence $M \in G(N)$. Define θ by

$$\theta(G[E']\tilde{Q}) = M$$

It will follow from the remarks preceeding Theorem 4 that θ is well defined, once we have checked that necessarily $f'(Z^h) = FZ(M)$ in the representation for M. For a prime q which does not divide t one has certainly

$$f'_q(Z_q^h) = FZ(M)_q$$

and if p divides t

$$f'_p(Z_p^h) = FZ(M)_p$$

as can readily be seen from the fact that $f(Z^h) = FZ(N)$ and using $f_t\alpha = \gamma f'_t$.

θ is one-one since, as we have already observed, $M \cong N$ would imply that $G[E']\tilde{Q} = G[E]\tilde{Q}$.

To see that θ is onto, choose $M \in G(N)$ represented by

$$E'' : Z^h \xrightarrow{f''} M \longrightarrow Q$$

with $f''(Z^h) = FZ(M)$. Then there are commutative diagrams

$$\begin{array}{ccccccc} E''_p & : & Z_p^h & \longrightarrow & M_p & \longrightarrow & Q_p \\ & & \alpha(p)\downarrow\cong & & \cong\downarrow & & \cong\downarrow\beta(p) \\ E_p & : & Z_p^h & \longrightarrow & N_p & \longrightarrow & Q_p \end{array}$$

since $M \in G(N)$. Hence $\alpha(p)_*[E'']_p = \beta(p)^*[E]_p$. Notice that

$$H^2(Q;Z^h) = \underset{p/t}{\oplus} H^2(Q_p;Z_p^h)$$

It remains to show that $[E'']$ lies in the orbit of $[E]$ under the $G(t) - \tilde{Q}$ action. Let $\beta = \pi\beta(p) : Q \to Q$ be the automorphism with $\beta_p = \beta(p)$. Recall that

$$H^2(Q;Z_t^h) \cong \text{char } Q^h \cong Q_{ab}^h$$

has exponent t. Hence the $G(t) = GL(h,Z_t)$ action factors through $GL(h,Z/t)$. Similarly, the $GL(h,Z_p)$ action on $H^2(Q_p;Z_p^h)$ factors through $GL(h,Z/p^{k(t)})$, where $p^{k(t)}$ denotes the highest power of p which divides t. Since

$$GL(h,Z/t) = \underset{p/t}{\Pi} GL(h,Z/p^{k(t)})$$

it follows that we can find $\alpha \in GL(h,Z_t)$ with

$$\alpha \equiv \alpha(p) \bmod p^{k(t)}$$

for all primes p which divide t, which implies that

$$\alpha_*[E''] = \beta^*[E]$$

and hence

$$G(t)[E'']\tilde{Q} = G(t)[E]\tilde{Q} .$$

It follows that $G[E'']\tilde{Q} \in \sigma^{-1}(G(t)[E]\tilde{Q})$ is such that

$$\theta(G[E'']\tilde{Q}) = M .$$

Hence θ is onto, and the proof of the theorem is finished.

We have seen that the (transitive) $GL(h,Z_t)$ action on $\sigma^{-1}(G(t)[E]\tilde{Q})$ factors through $GL(h,Z/t)$. Obviously, the image of $GL(h,Z) \to GL(h,Z/t)$ lies in the isotropy subgroup of any point of the orbit $\sigma^{-1}(G(t)[E]\tilde{Q}) \subset H^2(Q;Z^h)$. Thus, there is a surjective map of coker $(GL(h,Z) \to GL(h,Z/t)) \cong (Z/t)^*/\{\pm 1\}$ onto $G(N)$:

$$\delta(N) : (Z/t)^*/\{\pm 1\} \twoheadrightarrow G(N) .$$

From this we get an upper bound for the cardinality of $G(N)$.

Corollary 2 If N is a finitely generated nilpotent group with finite commutator subgroup, and if $t = \exp(Q(N)_{ab})$, then

$$|G(N)| \leq \varphi(t)/2$$

where φ denotes the Euler φ-function.

This is clear, since $(Z/t)^*$ has order $\varphi(t)$.

Notice that $\exp(Q(N)_{ab}) = \exp(Q(N^k)_{ab})$ for any $k \geq 1$, since $Q(N^k)_{ab} \cong (Q(N)^k)_{ab} \cong (Q(N)_{ab})^k$. This enables us to get some information about powers of members of the same genus.

Theorem 5 Let N be a finitely generated nilpotent group with finite commutator subgroup, and let $t = \exp(Q(N)_{ab})$. Then there is a commutative diagram

$$\begin{array}{ccc} (Z/t)^*/\{\pm 1\} & \xrightarrow{\delta(N)} & G(N) \\ \text{k-power} \downarrow & & \downarrow \text{k-power (i.e. } M \mapsto M^k) \\ (Z/t)^*/\{\pm 1\} & \xrightarrow{\delta(N^k)} & G(N^k) \end{array}$$

Proof. Notice that, using the Künneth-Theorem, one can identify $H^2(Q^k;(Z^h)^k)$ with the additive group of $k \times k$-matrices with entries in $H^2(Q;Z^h)$. Let $\bar{x} \in (Z/t)^*/\{\pm 1\}$ be represented by $x \in (Z/t)^*$. Then $\delta(N)\bar{x} = M$, where M may be represented by $X_*[E] \in H^2(Q;Z^h)$ with $X \in GL(h,Z_t)$ having determinant $\pm x$ mod t, and $[E]$ representing N. Thus $(\delta(N)\bar{x})^k = M^k$ is representable by a diagonal matrix $\mathrm{diag}(X_*[E],\ldots,X_*[E]) \in H^2(Q^k;(Z^h)^k)$. Since $\mathrm{diag}(X_*[E],\ldots,X_*[E]) = \mathrm{diag}(X,\ldots,X)_*\,\mathrm{diag}([E],\ldots,[E])$ and $\mathrm{diag}(X,\ldots,X) \in GL(hk,Z_t)$ has determinant $\pm x^k$ mod t, and since $\mathrm{diag}([E],\ldots,[E]) \in H^2(Q^k;Z^{hk})$ represents N^k, it follows that

$$\delta(N^k)(\bar{x}^k) = (\delta(N)\bar{x})^k$$

As an application of Theorem 5, we get the following result.

Theorem 6 Let N be a finitely generated nilpotent group with finite commutator subgroup, and let $t = \exp(Q(N)_{ab})$. Further let $M \in G(N)$. Then

$$N^{\varphi(t)/2} \cong M^{\varphi(t)/2}$$

where φ denotes the Euler φ-function.

Proof. Let $x \in (\mathbb{Z}/t)^*/\{\pm 1\}$ be such that $\delta(N)x = M$. Then $(\delta(N)x)^{\varphi(t)/2} = M^{\varphi(t)/2} = \delta(N^{\varphi(t)/2})(x^{\varphi(t)/2}) = \delta(N^{\varphi(t)/2})(1) = N^{\varphi(t)/2}$

Proof of Theorem 3. Let N(p) be as in Theorem 3. Then $M \in G(N(p))$ implies that $\mathbb{Z}/p \cong [M,M] \subset Z(M)$, since $\mathbb{Z}/p \cong [N(p), N(p)] \subset Z(N(p))$. Hence, there is a central extension

$$F : \mathbb{Z}/p \to M \to \mathbb{Z}^{p-1}$$

and

$$[F] \in H^2(\mathbb{Z}^{p-1}; \mathbb{Z}/p) \cong \Lambda^2 \operatorname{Hom}(H_1(\mathbb{Z}^{p-1}; \mathbb{Z}/p), \mathbb{Z}/p)$$

We will consider therefore [F] as an alternating form on $H_1(\mathbb{Z}^{p-1};\mathbb{Z}/p)$. Notice that a form $x \in H^2(\mathbb{Z}^{p-1};\mathbb{Z}/p)$ is non degenerate iff

$$x^{(p-1)/2} \in H^{p-1}(\mathbb{Z}^{p-1}; \mathbb{Z}/p)$$

is non zero. Since $[E(p)] = \Sigma x_i y_i$ is non degenerate, $[E(p)]^{(p-1)/2}$ generates $H^{p-1}(\mathbb{Z}^{p-1}; \mathbb{Z}/p) \cong \mathbb{Z}/p$.

Let $\alpha : N(p)_p \to M_p$ be an isomorphism. Then one has a commutative diagram

$$\begin{array}{ccccccc} E(p)_p & : & \mathbb{Z}/p & \to & N(p)_p & \to & \mathbb{Z}_p^{p-1} \\ & & b \downarrow \cong & & \cong \downarrow & & \psi \downarrow \cong \\ F_p & : & \mathbb{Z}/p & \to & M_p & \to & \mathbb{Z}_p^{p-1} \end{array}$$

from which we deduce that

$$\psi^*[F]_p = b[E(p)]_p$$

and hence

$$(*) \qquad \det \psi \cdot [F]_p^{(p-1)/2} = b^{(p-1)/2} \cdot [E(p)]_p^{(p-1)/2}$$

Because the canonical map $Z^{p-1} \to Z_p^{p-1}$ induces an isomorphism

$$H^*(Z_p^{p-1}; Z/p) \overset{\cong}{\to} H^*(Z^{p-1}; Z/p)$$

and because $b \in (Z/p)^*$, we deduce from equation (*) that

$$[F]^{(p-1)/2} \neq 0$$

or, the alternating form $[F]$ representing $M \in G(N(p))$ is non degenerate. Define

$$\varrho : G(N(p)) \to (Z/p)^*/\{\pm 1\}$$
$$M \mapsto \pm a$$

where a is such that

$$[F]^{(p-1)/2} = a[E(p)]^{(p-1)/2} \quad , \; a \in (Z/p)^*$$

We have to check that $\pm a$ depends only upon the isomorphism class of M. For this, let $\overline{M}$ be isomorphic to M, represented by

$$\overline{F} : Z/p \to \overline{M} \to Z^{p-1}$$

Then these exists a commutative diagram

$$\begin{array}{ccccc} \overline{F} : Z/p & \longrightarrow & \overline{M} & \longrightarrow & Z^{p-1} \\ c \Big\downarrow \cong & & \Big\downarrow \cong & & \varphi \Big\downarrow \cong \\ F : Z/p & \longrightarrow & M & \longrightarrow & Z^{p-1} \end{array}$$

from which we get $\varphi^*[F] = c[\overline{F}]$, and hence

$$\det \varphi \cdot [F]^{(p-1)/2} = c^{(p-1)/2}[\overline{F}] = \pm [\overline{F}]$$

or

$$[\overline{F}]^{(p-1)/2} = \pm [F]^{(p-1)/2} = \pm a \; [E(p)]^{(p-1)/2} ,$$

which shows that ϱ is well defined.

To see that ϱ is surjective, we choose $\pm a \in (Z/p)^*/\{\pm 1\}$ and consider

$$[E'] = ax_1y_1 + \sum_{i \geq 2} x_iy_i \quad .$$

Then

$$[E']^{(p-1)/2} = (\tfrac{p-1}{2})! \; a \; \Pi \; x_iy_i = a[E]^{(p-1)/2}$$

It remains to check that

$$E' : \mathbb{Z}/p \to M \to \mathbb{Z}^{p-1}$$

is an extension representing a group $M \in G(N(p))$. Clearly, $M_q \cong N(p)_q$ for $q \neq p$. Further, we notice that reduction mod p induces a surjection

$$red_p : GL(p-1, \mathbb{Z}_p) \twoheadrightarrow GL(p-1, \mathbb{Z}/p) .$$

Choose $\psi \in GL(p-1,\mathbb{Z}_p)$ such that

$$red_p \ \psi = diag(a^{-1},1,1,\ldots,1)$$

as automorphism of $H^1(\mathbb{Z}_p^{p-1};\mathbb{Z}/p) \cong H^1(\mathbb{Z}^{p-1};\mathbb{Z}/p)$, with respect to the basis $\{\bar{x}_1,\bar{x}_2,\ldots\bar{x}_{(p-1)/2}, \bar{y}_1,\bar{y}_2,\ldots,\bar{y}_{(p-1)/2}\}$.

Then

$$\psi^*[E']_p = [E]_p$$

and we conclude that there is a commutative diagram

$$\begin{array}{ccccc} \mathbb{Z}/p & \to & N(p)_p & \to & \mathbb{Z}_p^{p-1} \\ {\scriptstyle 1}\downarrow & & \downarrow & & \downarrow{\scriptstyle \psi} \\ \mathbb{Z}/p & \to & M_p & \to & \mathbb{Z}_p^{p-1} \end{array}$$

As ψ is an isomorphism, $N(p)_p \cong M_p$ and hence $M \in G(N(p))$. By construction

$$\varrho(M) = \pm a$$

To see that ϱ is injective, consider

$$F : \mathbb{Z}/p \to M \to \mathbb{Z}^{p-1} \ ; \quad \bar{F} : \mathbb{Z}/p \to \bar{M} \to \mathbb{Z}^{p-1}$$

with $M,\bar{M} \in G(N(p))$. Since both, $[F]$ and $[\bar{F}]$ are non degenerate alternating forms on $H_1(\mathbb{Z}^{p-1};\mathbb{Z}/p)$ we infer, by the fundamental theorem on alternating forms, that there exist $a,b \in (\mathbb{Z}/p)^*$ and automorphisms $\varphi,\psi \in GL(p-1,\mathbb{Z})$, such that

$$\varphi^*[F] = ax_1y_1 + \sum_{i\geq 2} x_iy_i \ ; \quad \psi^*[\bar{F}] = bx_1y_1 + \sum_{i\geq 2} x_iy_i \ .$$

But this implies that

$$\varrho(M) = \pm a \quad ; \quad \varrho(\bar{M}) = \pm b$$

Therefore, if $\varrho(M) = \varrho(\bar{M})$, then there exist $X \in GL(p-1,\mathbb{Z})$ with

$$X^*[F] = [\bar{F}] .$$

Of course, this last equation implies that $M \cong \bar{M}$, and therefore ϱ is one-one. Theorem 3 now follows, because the cardinality of $(\mathbb{Z}/p)^*/\{\pm 1\}$ is $(p-1)/2$.

For the further study of the interrelationship between genus and cancellation, we suggest the following problems.

1) Find a finitely generated <u>torsionfree</u> nilpotent group N with $|G(N)| > 1$.

2) Prove that $N^k \cong M^k$ implies $G(N) = G(M)$, at least for N and M finitely generated nilpotent groups with finite commutator subgroups.

3) Find a "genuine" non-cancellation example; i.e. find finitely generated nilpotent groups M,N and L with

$$M \times L \cong N \times L \quad \text{but} \quad G(N) \neq G(M) .$$

References

[1] A.K. Bousfield and D.M. Kan, Homotopy limits, completions and localizations, Lecture Notes in Mathematics 304, Springer Verlag (1972).

[2] P.J. Hilton, Localization and cohomology of nilpotent groups, Math. Zeits. 132 (1973), 263-286.

[3] P.J. Hilton, G. Mislin and J. Roitberg, H-spaces of rank two and non-cancellation phenomena, Inv. Math. 16 (1972), 325-334.

[4] P.J. Hilton, G. Mislin and J. Roitberg, Localization theory for nilpotent groups and spaces, Notas de Mathematica, North Holland (to appear).

[5] M. Lazard, Sur les groupes nilpotents et les anneaux de Lie, Ann. Sci. Ecole Norm. Sup. (3) 71 (1954), 101-190.

[6] A.I. Malcev, On a class of homogeneous spaces, Izv. Akad. Nauk. SSSR Ser. Mat. 13 (1949), 9-32; English transl., Amer. Math. Soc. Transl. (1) 9 (1962), 276-307.

[7] J.W. Milnor, private communication.

[8] G. Mislin, The genus of an H-space, Lecture Notes in Mathematics 249, Springer Verlag (1971), 75-83.

[9] G. Mislin, Cancellation properties of H-spaces, Comment. Math. Helv. (to appear).

[10] P.F. Pickel, Finitely generated nilpotent groups with isomorphic finite quotients, Trans. Amer. Math. Soc. 160 (1971), 327-341.

[11] U. Stammbach, Homology in group theory, Lecture Notes in Mathematics 359, Springer Verlag 1973.

[12] R.B. Warfield, Lecture notes on nilpotent groups, Forschungsinstitut für Mathematik, ETH Zürich, 1973.

[13] C. Wilkerson, Genus and cancellation, preprint.

[14] A. Zabrodsky, On the genus of finite CW H-spaces. Comment. Math. Helv. (to appear).

LIE GROUPS FROM A HOMOTOPY POINT OF VIEW

by

David Rector and James Stasheff

Rice University, Houston

Temple University, Philadelphia

One of the most beautiful and significant parts of mathematics is the theory of Lie groups. It can be regarded as a part of algebra _or_ analysis _or_ geometry _or_ topology with major connections to number theory and even probability, crystallography, and physics. Although many results in the field are proved using the techniques of algebra or analysis, they are often seen to depend in some poorly understood way only on the underlying homotopy type. This is particularly true where the proof depends on a case by case checking using Cartan's complete classification of Lie groups. An outstanding example of this (phenomenon) is:

[Baum and Browder; Scheerer] Compact Lie groups are isomorphic if and only if they have the same homotopy type.

Indeed this result says the entire theory of semi-simple Lie groups must in principle be homotopy invariant.

To pursue a better understanding of Lie groups from a homotopy point of view, we extend our horizons to a larger class of groups which by very definition will be characterized in homotopy language. We immediately encounter new examples:

1) new group structures on the homotopy type e.g. S^3 [Slifker]
2) new group extensions, e.g. of S^3 by S^3 [Terrell]
3) new underlying homotopy types e.g. the [Hilton-Roitberg] criminal which is an S^3-bundle E_5 over S^7 induced as a space from $Sp(2)$ by a map of degree 5 of S^7 onto itself.

Thus, Hilbert's Fifth Problem has a negative solution in the homotopy category: _there is a topological group of the homotopy type of the compact manifold_ E_5 _which is not of the homotopy type of a Lie group_. However this and so far all the new discoveries of non-Lie examples do belong to the genuses of Lie groups, i.e.

for each prime p including 0 they are p-equivalent to Lie groups.

Thus, it is appropriate to bring to bear the techniques of localization and completion. The p-local homotopy theory of finite H-spaces has several nice features, particularly p-universality and unique factorization of the underlying homotopy type (see Wilkerson's talk). Ideally we might hope for a p-local theory analogous to the classical theory--but in homotopy terms--and some sort of assembly theory which would account for Lie groups in terms of compatibility of the p-local pieces. The p-local picture will however look somewhat different from the classical case, as is revealed by spheres.

The only spheres which are groups are S^1 and S^3. Rationally, every odd dimensional sphere is a group and [Hopf] showed essentially that every Lie group is rationally a product of odd dimensional sphere, $\prod_{i=1}^{r} S^{2n_i-1}$. The set $\{n_i\}$ is called the <u>type</u> of the group and r is the <u>rank</u>. [Sullivan] has shown that for odd p, a sphere S^n is p-equivalent to a group precisely when $n = 2d - 1$ where $d | p - 1$. Results of [Serre] and [Kumpel] show that a Lie group is p-equivalent to a product of odd dimensional spheres precisely if p is greater than the largest n_i in the type. Note that here again the proof involves case by case checking using special techniques in each case.

The approach we have in mind is well illustrated by the homotopy theoretic statement of the following results:

1) [Hopf] <u>If</u> H <u>is a finite</u> H-<u>complex, then</u> H <u>is rationally the product of odd dimensional spheres and the type is a homotopy invariant</u>.

2) [Wilkerson; Underwood] <u>If</u> H <u>is a finite complex having the homotopy type of a group, then</u> H <u>is</u> p-<u>equivalent to a product of odd spheres iff</u>

a) H <u>is</u> p-<u>torsion free</u>

b) $p > \underline{\max}\, n_i$, $n_i \in \underline{\text{type}}\ H$.

The proofs of these facts are of course general since, so far, no classification theorem for H-complexes is known. The proof of Hopf, in particular, led to important algebraic ideas (Hopf algebras).

We turn now to a list of properties of Lie groups which are nicely expressible in homotopy language. Properties 1 through 4 are progressively more restrictive as far as is known.

1. G is a finite complex and H-space, in fact of the homotopy type of a loop space ΩBG.
2. G has a maximal torus T.
3. The normalizer of T in G is an extension of T by a finite group W, the Weyl group.
4. If $H^*(BG;k)$ is p-torsion free, $p = \text{char } k$, it is isomorphic to the invariant elements $H^*(BT)^W$.
5. If G is compact, G is faithfully represented in some $U(n)$.
6. The loop space ΩG is torsion free.
7. Classification: If G is compact and simply connected, then G may be written uniquely as a product of simple Lie groups: namely the classical groups
   $SU(n)$ $n \geq 2$
   $Sp(n)$ $n \geq 2$
   $Spin(n)$ $n \geq 5$
   or the exceptional groups G_2, F_4, E_6, E_7, E_8.

Therefore we call such groups semi-simple.

§1. Just the assumption that G is a finite complex and H-space enabled Hopf to establish its rational homotopy type. It is also sufficient for the following:

[Browder] $\pi_2(G) = 0$

[Browder] G satisfies Poincaré duality.

Now G has the homotopy type of a topological group if and only if G has the homotopy type of the loops on the classifying space BG. (The group structure on G corresponds in a very strong sense to the loop space structure.)

Definition. A finite loop space is a topological group G of the homotopy type of a finite complex. We refer to the classifying space BG as a finite loop structure. Two are said to be equivalent if the classifying spaces are homotopy equivalent.

§2. A compact connected abelian Lie group is a torus $T^r = S^1 \times \ldots \times X^1$ where r denotes the number of factors. A compact Lie group G contains a maximal such subgroup and r then equals the rank of G. In homotopy terms, a homomorphism $h: H \to G$ corresponds to a map $Bh: BH \to BG$. For Lie groups with h being the inclusion of a subgroup, G/H is again a finite dimensional manifold of the homotopy type of the fibre of Bh. This motivates the following:

Definition. [Rector]. For a finite loop space G, a sub-finite loop space is a map $f: BH \to BG$ with homotopy theoretic fibres of the homotopy type of a finite complex, denoted G/H.

Up to homotopy $\Omega BH \to \Omega BG \to G/H$ is a principal fibration. This is used to prove

Theorem. [Rector]. G/H satisfies Poincaré duality.

Definition. [Rector]. A maximal torus in a finite loop space G is a sub-finite loop space $f: BT^r \to BG$ where r equals the rank of G.

Not all finite loop spaces have maximal tori [Rector].

§3. For Lie groups, the Weyl group W is defined as the quotient N/T where N is the normalizer of T in G. As such, W acts on T by inner automorphisms. The order of W is the product of the numbers n_i in the type of G.

Theorem. [Rector]. For a compact Lie group G and a maximal torus $i: T \subset G$, the Weyl group W is isomorphic to the set of homotopy classes of maps $\alpha: BT \to BT$ such that $(B_i) \circ \alpha \simeq B_i$.

This suggests:

Definition. [Rector]. For a maximal torus $f: BT \to BG$, the the Weyl group

$$W(f) = \{\alpha \in [BT,BT] \mid f \circ \alpha \simeq f\}.$$

It is not known if $W(f)$ really depends on f. It is however always finite [Rector].

For Lie groups, the action of W on T does not determine N as a group extension, though the remaining data needed seem to be in the 2-torsion of the cohomology of W. Perhaps at all odd primes the action determines N. In any case, Curtis, Williams and Wiederhold have recently shown that N determines G, i.e. *compact connected semisimple Lie groups* G_1, G_2 *are isomorphic if and only if their corresponding normalizers* N_1 *and* N_2 *are isomorphic*.

It is worth noting that although this is a result in classical Lie theory proved by classical techniques, its discovery was motivated by the homotopy theoretic approach.

For a Lie group G and a maximal torus T, Bott shows G/T is torsion free in cohomology. The Euler characteristic of G/T is the order of the Weyl group. The situation for finite loop spaces is unknown, except that $\chi(G/T)$ is an upper bound for the order of W.

§4. There are many important consequences of the lack of p-torsion in $H_*(G)$ or equivalently in $H^*(BG)$.

[Borel]. For a Lie group G and subgroup H of the same rank, both without p-torsion, there is an isomorphism with coefficients in a field of characteristic p, a prime or zero:

$$H^*(G/H) \approx H^*(BH)//H^*(BG).$$

A proof due to [Baum] works well for finite loop spaces in full generality. With real coefficients, Borel had a similar result for any sub Lie group H. Recent proofs in the p-torsion free case, due to [Gugenheim-May], [Munkholm] or [Husemoller-Moore-Stasheff], are sufficiently general to yield:

For a finite loop space G *and a sub-finite loop space* H, *both without* p-*torsion, then, provided* H *has a maximal torus* $BT \to BH$ *there is an isomorphism with coefficients in a field* F *of characteristic* p:

$$H^*(G/H) \approx \mathrm{Tor}_{H^*(BG)}(H^*BH;F).$$

In particular in the absence of p-torsion, if G has a maximal torus, $H^*(BG)$ is faithfully represented in the invariant sub-algebra $H^*(BT)^W$, but we do not know if the map is onto. However, Wilkerson has shown: <u>If</u> $H^*(BG;Q) \approx H^*(BT;Q)^W$, <u>then</u> G <u>has the type of a Lie group</u>.

In the presence of p-torsion, the picture is very obscure. One of the few conjectures or rather questions available is: For a finite loop space, is $H^*(BG)$ at least Noetherian?

§5. The Peter-Weyl theorem asserts that any compact Lie group can be regarded as a subgroup of a unitary group $U(n)$. To our knowledge, there has been no attack on the corresponding result for finite loop spaces. A major tool of Lie group theory involves the "weights" of representations. From a homotopy point of view, these are best expressed in terms of the cohomology map of

$$T \to G \to U(n)$$

if T is a maximal torus.

More generally one would like to understand the representation theory of Lie groups in terms of all the maps $BG \to BU(n)$. Initial results have been provided by [Hubbuck, Mahmud and Adams], but again case by case analysis is involved.

§6. Bott showed ΩG was torsion free for a compact, connected Lie group G. This has proved to be a very useful additional assumption but an extremely difficult question for finite loop spaces. [Kane] has recently shown that for a simply connected finite H-space G, if ΩG is p-torsion free, then G is p^2-torsion free.

§7. If there is a classification theorem for finite loop spaces localized at a prime p, the list will have to be much longer. For Lie groups, the Weyl groups are all reflection groups. By considering a p-adic torus, [Sullivan, Clark, Ewing and Wilkerson] have used "pseudo-reflection groups", e.g. $\Sigma_n \int Z/p-1$ to produce many new finite p-local loop spaces. Some of these occur as factors of localized Lie groups (see Stasheff's talk). Again, in the presence of p-torsion very little is known (see Harper's talk).

There are the bare beginnings of a classification theory, but nothing like a complete description is available. If the cohomology algebra of the finite loop space is suitable restricted, there are some partial results:

[Hubbuck-Ewing]. A finite loop space of rank ≤ 6 has the type of a Lie group.

[Wilkerson, Underwood]. A finite loop space G is mod p equivalent to a product of spheres if G is p-torsion free and $p > \max n_i \in$ type G.

[Harper-Wilkerson-Zabrodsky]. A finite loop space G is mod p equivalent to a product of spheres and sphere bundles over spheres if $2p > \max n_i$ or $P^1 x_i = \lambda x_j$ and $\max (n_i - n_j) < 2(p-1)$.

[Rector]. For loop structure B on S^3 of the genus of $HP(\infty)$, there is an invariant $\frac{B}{p}$, an element of $\{\pm 1\}$ such that the set $\{\frac{B}{p}, p \text{ prime}\}$ determines B and all possible combinations occur for some B.

In conclusion we remark that it seems to us there are now enough individual results and examples to indicate the relevance of the homotopy theoretic point of view. It is time to direct our efforts to the significant theoretical questions bearing on a better understanding of Lie groups and to the applications of this new understanding.

BIBLIOGRAPHY

There follows a list of books and papers related to the topology of Lie groups and their homotopy-theoretic study. This list is by no means complete. Extensive bibliographies relating to the study of H-spaces may be found in the survey article by Curtis (see first reference below); in J. Stasheff, H-spaces from a homotopy point of view, Lecture Notes in Math. Vol. 161, Springer (1970); and in H-spaces Neuchâtel (Suisse) Août 1970, Lecture Notes in Math. Vol. 196, Springer (1970).

1. J. F. Adams, Lectures on Lie Groups, Benjamin, New York (1969).
2. __________, "The sphere considered as an H-space mod p", Quart. J. Math. (Oxford), 12 (1961), 52-60.

3. P. F. Baum, "On the cohomology of homogeneous spaces", Topology 7 (1968), 15-38.

4. P. F. Baum and W. Browder, "The cohomology of quotients of classical groups", Topology 3 (1965), 305-336.

5. A. Borel, Cohomologie des espaces localement compact d'apres J. Leray, Lecture Notes in Math. Vol. 2, Springer (1964).

6. ________, Linear Algebraic Groups, Benjamin, New York (1969).

7. ________, "Sur le cohomologie des espaces fibrés principaux et des espaces homogènes de groupes de Lie compact", Ann. of Math., 57 (1953), 115-207.

8. A. Borel and F. Hirzebruch, "Characteristic classes and homogeneous spaces I, II, III", Amer. J. Math., 80 (1958), 458-538; 81 (1959), 315-382; 82 (1960), 491-504.

9. R. Bott and H. Samelson, "Applications of the theory of Morse to symmetric spaces", Amer. J. of Math., 80 (1958), 964-1029.

10. G. Bredon, Introduction to compact transformation groups, Academic Press, New York (1972).

11. W. Browder, "Fiberings of spheres which are rational homology spheres", Bull. A.M.S., 68 (1962), 202-203.

12. __________, "Torsion in H-spaces", Ann. of Math., 74 (1961), 24-51.

13. C. Chevalley, Séminaire C. Chevalley, 1956-1958, Classification des Groupes de Lie Algébriques, École Normade Supérieure, Paris.

14. ____________, Theory of Lie Groups, Princeton University Press, Princeton, New Jersey (1946).

15. A. Clark, "On π_3 of finite dimensional H-spaces", Ann. of Math., 78 (1963), 193-196.

16. Morton Curtis, "Finite dimensional H-spaces", Bull. A.M.S., 77 (1971), 1-12.

17. ____________, "H-spaces mod p (II)", H-spaces Neuchâtel (Suisse) Août 1970, Lecture Notes in Math., Vol. 196, Springer (1971), 11-19.

18. Roy R. Douglas and Francois Sigrist, "Sphere Bundles over spheres and H-spaces", Topology, 8 (1969), 115-118.

19. J. Ewing, "The non-splitting of Lie groups as loop spaces", (preprint).

20. ______, "On the type of associative H-spaces", preprint, Aarhus Universitet 1970-71, no. 15.

21. V. K. A. M. Gugenheim and J. P. May, On the theory and applications of differential torsion products, Memoirs A.M.S. 142 (1974).

22. P. Hilton, G. Mislin and J. Roitberg, Sphere bundles over spheres and non-cancellation phenomena, J. Lond. Math. Soc., (2) 6 (1972), 15-23.

23. ______, H-space of rank two and non-cancellation phenomena, Inv. Math. 16 (1972), 325-334.

24. P. Hilton and J. Roitberg, "On principal S^3-bundles over spheres", Ann. of Math., 90 (1969), 91-107.

25. ______, "On the classification problem for H-spaces of rank two, Comm. Math. Helv. 46 (1971), 506-516.

26. H. Hopf, "Uber die Topologie der Gruppen-Mannigfaltigkeiten und Ihre Verallgemeinerungen", Ann. of Math., 42 (1941), 22-52.

27. L. Hodgkin, "On the K-theory of Lie groups", Topology, 6 (1976), 1-36.

28. J. R. Hubbuck, "Generalized cohomology operations and H-spaces of low rank", Trans. A.M.S., 141 (1969), 335-360.

29. ______, "On homotopy commutative H-spaces", Topology, 8 (1969), 119-126.

30. ______, "S. H. M. self maps of the classical Lie groups", (preprint).

31. S. Y. Husseini, "The topology of classical groups and related topics", Gordon and Breach, New York, 1969.

32. D. Husemoller, J. Moore, and J. Stasheff, "Differential homological algebra and homogeneous spaces", J. of Pure and Applied Algebra, to appear.

33. L. Illusie, "Travaux de Quillen sur la cohomologie des groupes", Séminaire Bourbaki, 405, Lecture Notes in Math. Vol. 317, Springer.

34. P. G. Kumpel, Jr., "Lie groups and products of spheres", Proc. A.M.S., 16 (1965), 1350-6.

35. M. Mimura and H. Toda, "Cohomology operations and the homotopy of compact Lie groups, I", Topology, 9 (1970), 317-336; II, to appear.

36. C. E. Miller, "The topology of rotation groups", Ann. of Math., 57 (1953),

90-114.

37. Guido Mislin, "H-spaces mod p (I)", H-spaces Neuchâtel (Suisse) Août 1970, Lecture Notes in Math., Vol. 196, Springer (1971).

38. H. Munkholm, "A collapse result for the Eilenberg-Moore spectral sequence", Bull. A.M.S., 79 (1973), 115-118.

39. Goro Nishida, "On a result of Sullivan and the mod-p decomposition of Lie groups", (preprint).

40. S. Ochiai, "On the type of an associative H-space", Proc. Jap. Acad., 45 (1969), 92-94.

41. D. G. Quillen, "The Adams conjecture", Topology, 10 (1971), 67-80.

42. __________, "On the cohomology and K-theory of the generalized linear groups over a finite field", Ann. of Math., 96 (1972), 552-586.

43. D. L. Rector, "Loop structures on the homotopy type of S^3", Symposium on Algebraic Topology, Battelle, 1971, Lecture Notes in Math., Vol. 249, Springer (1971), 99-105.

44. __________, "Subgroups of finite dimensional topological groups", J. of Pure and Appl. Algebra, 1 (1971), 253-273.

45. J. P. Serre, Algèbres de Lie semi-simples complexes, Benjamin, New York (1966).

46. __________, Groupes d'homotopie et classes de groupes abéliens, Ann. of Math., 58 (1953), 258-294.

47. __________, Lie algebras and Lie groups, Benjamin, New York (1965).

48. H. Scheerer, "Homotopie-äquivalente Kompakte Liesche Gruppen", Topologie, 7 (1968), 227-232.

49. J. F. Slifker, "Exotic multiplications on S^3", Quart. J. Math. Oxford, 16 (1965), 322-359.

50. J. D. Stasheff, "Manifolds of the homotopy type of (non-Lie) groups", Bull. A.M.S., 75 (1969), 998-1000.

51. D. Sullivan, Localization, periodicity and Galois symmetry, Geometric Topology, I, Notes, M.I.T., (1970), revised 1971.

52. George Terrell, Thesis, Rice University, in preparation.

53. R. Underwood, "Primes which are regular for associative H-spaces", Bull.

A.M.S., 79 (1973), 493-496.

54. ___________, Finite dimensional associative H-spaces and products of spheres, Trans. A.M.S., to appear.

55. H. Weyl, Classical Groups, Princeton University Press, Princeton, New Jersey (1946).

56. S. Weingram, "On the incompressibility of certain maps", (preprint).

57. C. Wilkerson, "Genus and cancellation", to appear.

58. ___________, "K-theory operations in mod p loop spaces", (preprint).

59. ___________, "Rational maximal tori", to appear.

60. C. Wilkerson and A. Zabrodsky, "Mod p decompositions of mod p H-spaces", (preprint).

61. I. Yokota, "On the cellular decompositions of unitary groups", J. Inst. Polytech. Osaka City U., A7 (1956), 39-49.

NILPOTENT GROUPS, HOMOTOPY TYPES AND RATIONAL LIE ALGEBRAS

Joseph Roitberg
Institute for Advanced Study, Princeton, New Jersey
and
Battelle Research Center, Seattle, Washington

§1. In this talk, we discuss some aspects of the relationship between the category $\mathcal{N}$ of nilpotent groups and the homotopy category $\mathcal{H}$ of 1-connected CW-spaces. That such a relationship exists and is fruitful has been illustrated in recent years in the work of Bousfield-Kan ([2], [3], [4]) and Hilton-Mislin-Roitberg ([8], [9], [10]). (See also [20].) Mislin's talk at this conference ([17]), in which it is demonstrated that the category $\mathcal{N}$ exhibits non-cancellation phenomena, much in the same way as with the category $\mathcal{H}$ (see [11]), provides further justification for the feeling that the "correct" group theory for 1-connected homotopy theory is nilpotent group theory.

The point of view we adopt here is due to Quillen [19] and consists of identifying the "rationalizations" $\mathcal{N}_0, \mathcal{H}_0$ of the categories $\mathcal{N}, \mathcal{H}$ with appropriate categories of rational Lie algebras (precise definitions below). We are, in this way led to make certain conjectures about the category $\mathcal{H}$ which are analogs of known facts about the category $\mathcal{N}$. All this is discussed in §2.

It is also possible to work the other way around, taking known phenomena in $\mathcal{H}$ and attempting to formulate analogs in $\mathcal{N}$. An illustration of this sort is the Mimura-Toda phenomenon ([15]), an analog of which was shown to me by John Milnor, in response to my question. Milnor's proof of this group-theoretic analog makes essential use of the connection between nilpotent groups and nilpotent Lie algebras over $\mathbb{Q}$ and led the author to try to study the Mimura-Toda phenomenon from a Lie-algebraic standpoint. §3 is devoted to giving the precise statements of Milnor's group-theoretic results as well as our homotopy-theoretic results, rederiving in a systematic and essentially algebraic manner the Mimura-Toda phenomenon. The proofs of all these results are, however, deferred to a separate publication ([21]).

§2. In addition to the categories $\mathcal{N}, \mathcal{H}$, we will be concerned with the following four categories:

$\mathcal{N}_0$ = the category of rational, that is torsion-free divisible, nilpotent groups;

$\mathcal{L}$ = the category of nilpotent Lie algebras over $\mathbb{Q}$;

$\mathcal{H}_0$ = the homotopy category of rational, 1-connected CW-spaces. (A 1-connected CW-space X is rational if its integral homology groups $H_i(X)$, $i > 0$, are rational vector spaces.);

$\mathcal{L}^*$ = the "homotopy" category of reduced, differential graded Lie algebras over $\mathbb{Q}$. (A graded Lie algebra L is reduced if its non-0 elements have positive degree.)

(To obtain $\mathcal{L}^*$, begin with the category of reduced, differential graded Lie algebras over $\mathbb{Q}$, consider the collection of all morphisms $\mu : L \to L'$ such that the induced homology map $H(\mu) : H(L) \to H(L')$ is an isomorphism, and then pass to the corresponding localization; see [19]. Our $\mathcal{L}^*$ is Quillen's more suggestive, but also more cumbersome, $\mathrm{Ho}(\mathrm{DGL})_1$.)

The following two fundamental theorems should justify the title of this paper. The proofs may be found in [19].

<u>Theorem 1.</u> <u>There is an equivalence of categories</u> $\mathcal{N}_0 \underset{\exp}{\overset{\log}{\rightleftarrows}} \mathcal{L}$. <u>Further, if</u> $N \in \mathcal{N}_0$ <u>and</u> $L \in \mathcal{L}$ <u>correspond under this equivalence, then there is a natural</u> $\mathbb{Q}$-<u>vector space isomorphism</u> $N_{ab} \cong L_{ab}$ <u>of the corresponding abelianizations.</u>

<u>Theorem 2.</u> <u>There is an equivalence of categories</u> $\mathcal{H}_0 \underset{\varepsilon}{\overset{\lambda}{\rightleftarrows}} \mathcal{L}^*$. <u>Further, if</u> $W \in \mathcal{H}_0$ <u>and</u> $L \in \mathcal{L}^*$ <u>correspond under this equivalence, then there is a natural graded Lie algebra isomorphism</u> $\prod(W) \cong H(L)$, <u>where</u> $\prod(W)$ <u>is the Whitehead product Lie algebra of</u> $W(\prod_i(W) = \pi_{i+1}(W))$.

Thinking of a nilpotent group G as an Eilenberg-MacLane space K(G, 1), it seems natural to expect a general theorem encompassing both Theorems 1 and 2 involving the category $\mathcal{NH}$ of nilpotent spaces ([4]). We state this formally as a problem.

<u>Problem.</u> Give a Lie-algebraic description of the category $\mathcal{NH}_0$.

More precisely, we conjecture an equivalence of categories $\mathcal{NH}_0 \rightleftarrows \mathcal{NL}^*$, where $\mathcal{NL}^*$ is defined as follows. An object $L \in \mathcal{NL}^*$ is, first of all, a differential, graded Lie algebra over $\mathbb{Q}$, all of whose non-0 elements have non-negative degree; furthermore, if $M = H(L)$, then the graded Lie algebra M is required to satisfy the following nilpotency condition: let $M(n) = \Gamma^1 M(n)$ denote the set of elements of M of degree n and inductively, let $\Gamma^{i+1}M(n)$ = subspace of $\Gamma^i M(n)$ spanned by all elements of the form $[a_0, a_n]$, $a_0 \in M(0)$, $a_n \in \Gamma^i M(n)$. Then, for each $n \geq 0$, there should exist $i_n \geq 1$ such that $\Gamma^{i_n} M(n) = 0$.

Remark. The proposed equivalence $\mathcal{NH}_0 \rightleftarrows \mathcal{NL}^*$ should be set up along the lines of Quillen's arguments proving Theorem 2. An important point in the proof of Theorem 2 is the use of E. Curtis' convergence theorem ([5]). Curtis' theorem was originally proved only for the category $\mathcal{H}$, but has been recently extended to the category $\mathcal{NH}$ by Bousfield-Kan [4].

We conclude this section with two conjectures.

Conjecture 1. If X is a finite, 1-connected CW-space, then Aut(X), the group of homotopy classes of homotopy equivalences of X with itself, is finitely presented.

Conjecture 2. If X is a finite, 1-connected CW-space, then $\mathcal{G}(X)$, the genus of X, is a finite set. (By the genus of X, we mean the set of all homotopy types of finite, 1-connected CW-spaces Y such that for each prime p, $Y_p \simeq X_p$; see [16], [10].)

Both these conjectures have group-theoretic analogs which are, in fact, theorems. Replacing X by a finitely generated nilpotent group G, we have G. Baumslag's theorem ([1]) that Aut(G) is finitely presented and Pickel's theorems ([18]) implying that $\mathcal{G}(G)$ (essentially the same definition as above: we take the set of all isomorphism classes of finitely generated nilpotent groups H such that for each prime p, $H_p \cong G_p$; see [10]) is a finite set. Both Baumslag's and Pickel's methods essentially involve the use of Theorem 1, so the hope is that Theorem 2 can be brought to bear on Conjectures 1 and 2.

We remark that some fragmentary results are available on Conjectures 1 and 2. With regard to Conjecture 1, the reader may wish to consult D. W. Kahn [12], [13].

§3. We say that a group homomorphism $\varphi : G \to H$ is a p-isomorphism ([7]), p a prime, if (a) $\ker\varphi$ consists of torsion prime to p and (b) given $y \in H$, there exist $x \in G$ and an integer n prime to p such that $y^n = \varphi(x)$. Similarly, we say that a continuous map $f : X \to Y$ is a p-equivalence if the induced integral homology maps $H_i(f) : H_i(X) \to H_i(Y)$, $i \geq 0$, are p-isomorphisms of groups.

The notion of p-equivalence of spaces goes back to Serre's fundamental paper on classes of abelian groups. Its study has been taken up again in recent years by Mimura-O'Neill-Toda [14] and Mimura-Toda [15]. Of particular interest to us here is the type of example constructed in [15], of which a version is stated in Theorem 4 below and which is at the origin of this work.

We now state the main results. For the proofs, we refer to [21].

Lemma 1. (J. Dyer [6]). There exists a finite-dimensional nilpotent Lie algebra L over $\mathbb{Q}$ with the property that any automorphism $\omega : L \to L$ is congruent, modulo $[L, L]$, to the identity.

Using Lemma 1 in conjunction with Theorem 1, we obtain

Theorem 3. (Milnor). There exist finitely generated nilpotent groups G, H and a p-isomorphism $\varphi : G \to H$ such that no map $\psi : H \to G$ is a p-isomorphism.

The graded analog of Lemma 1 is the following.

Lemma 2. There exists a reduced, differential graded Lie algebra L of finite type over $\mathbb{Q}$ with the property that any "weak" automorphism $\omega : L \to L$ (that is, the induced homology map $H(\omega) : H(L) \to H(L)$ is an automorphism) is congruent, modulo $[L, L]$, to the identity. (Indeed, in contrast with Lemma 1, it is not even necessary to work modulo $[L, L]$.)

L may furthermore be chosen to be of totally finite dimension.

Using Lemma 2 in conjunction with Theorem 2, we obtain

Theorem 4. There exist 1-connected CW-spaces of finite type X, Y and a p-equivalence $f : X \to Y$ such that no map $g : Y \to X$ is a p-equivalence.

X and Y may furthermore be chosen so that either (a) they both have only finitely many non-0 homotopy groups or (b) they both are finite complexes.

We stress the fact that our proof of Theorem 4 is quite algebraic and does not involve extensive homotopy-theoretic calculations, such as those in [15].

Remarks. (1) It is interesting to contrast Theorem 3 (and Theorem 4) with Theorem 6.8 of [7].

(2) If we do not restrict the groups G, H in Theorem 3 to be finitely

generated, then Peter Hilton has pointed out a simple construction of a "non-invertible" p-isomorphism $\varphi : G \to H$. In fact, let $\varphi : \mathbb{Z} \to \mathbb{Z}_p$ be the inclusion of the integers into the p-localized integers. Then φ is clearly a p-isomorphism but, since $\operatorname{Hom}(\mathbb{Z}_p, \mathbb{Z}) = 0$, there is no p-isomorphism $\psi : \mathbb{Z}_p \to \mathbb{Z}$. Similar remarks apply, of course, to Theorem 4.

(3) If G, H are finitely generated abelian groups, then it is easy to see that a p-isomorphism $\varphi : G \to H$ is, indeed, invertible. Similarly, if X, Y are finite CW-spaces and are both H-spaces (or co-H-spaces), then a p-equivalence $f : X \to Y$ is invertible; see [14].

Bibliography

[1] G. Baumslag, Lecture notes on nilpotent groups, A.M.S. Regional Conference Series No. 2(1971).

[2] A. K. Bousfield and D. M. Kan, Homotopy with respect to a ring, Proc. Symp. Pure Math., Amer. Math. Soc. 22(1971), 59-64.

[3] ______________, Localization and completion in homotopy theory, Bull. Amer. Math. Soc. 77(1971), 1006-1010.

[4] ______________, Homotopy limits, completions and localizations, Lecture Notes in Mathematics 304, Springer-Verlag (1972).

[5] E. B. Curtis, Some relations between homotopy and homology, Ann. of Math. 83(1965), 386-413.

[6] J. L. Dyer, A nilpotent Lie algebra with nilpotent automorphism group, Bull. Amer. Math. Soc. 76(1970), 52-56.

[7] P. J. Hilton, Localization and cohomology of nilpotent groups, Math. Zeit. 132(1973), 263-286.

[8] P. J. Hilton, G. Mislin and J. Roitberg, Topological localization and nilpotent groups, Bull. Amer. Math. Soc. 78(1972), 1060-1063.

[9] ______________, Homotopical localization, Proc. London Math. Soc. 26(1973), 693-706.

[10] ______________, Localization of nilpotent groups and spaces, (In preparation).

[11] P. J. Hilton and J. Roitberg, On principal S^3-bundles over spheres, Ann. of Math. 90(1969), 91-107.

[12] D. W. Kahn, The group of stable self-equivalences, Topology 11(1972), 133-140.

[13] ______, A note on H-equivalences, Pacific Jour. of Math. 42(1972), 77-80.

[14] M. Mimura, R. C. O'Neill and H. Toda, On p-equivalence in the sense of Serre, Japanese Jour. of Math. 40(1971), 1-10.

[15] M. Mimura and H. Toda, On p-equivalences and p-universal spaces, Comm. Math. Helv. 46(1971), 87-97.

[16] G. Mislin, The genus of an H-space, Lecture Notes in Mathematics 249, Springer-Verlag (1971), 75-83.

[17] ______, Nilpotent groups with finite commutator subgroups, Proc. this Conference, Lecture Notes in Mathematics, Springer-Verlag, (To appear).

[18] P. F. Pickel, Finitely generated nilpotent groups with isomorphic finite quotients, Trans. Amer. Math. Soc. 160(1971), 327-341.

[19] D. G. Quillen, Rational homotopy theory, Ann. of Math. 90(1969), 205-295.

[20] J. Roitberg, Note on nilpotent spaces and localization, Math. Zeit., (To appear).

[21] __________, Rational Lie algebras and p-isomorphisms of nilpotent groups and homotopy types, (To appear).

H-SPACE NEWSLETTER - MAY, 1974

JAMES STASHEFF

TEMPLE UNIVERSITY

In addition to the reports in these proceedings, there are several newsworthy items:

SMALL H-SPACES

Terrell (Rice) has exotic extensions of S^3 by S^3.

Let $S^{2n+1} \to B_n(p) \to S^{2n+2(p-1)+1}$ be the bundle with characteristic class α generating $\pi_{2n+2(p-1)}(S^{2n+1})_p$. Stasheff's report on Nishida's decomposition includes results of Mimura and Toda on the mod p equivalence of low rank Lie groups to products of spheres and $B_n(p)$'s.

Wilkerson-Zabrodsky: A simply connected H-space X with $H^*(X;Z/p) \cong E(x_1,\ldots,x_n)$ with $\deg x_i = 2r_i - 1$ $r_i \leq r_{i+1}$ and for each i, $\mathcal{O}^1 x_i = \lambda x_j$ some j is mod p equivalent to a product of spheres and $B_n(p)$'s provided $r_n - r_1 < 2(p-1)$.

Harper: The same conclusion holds if the hypothesis for r is replaced by "$r_n < 2p$ and X rationally primitively generated."

Wilkerson/Ewing: $B_1(p)$ and $B_{p-2}(p)$ are loop spaces.

Ewing: So are $B_7(17)$, $B_5(19)$, $B_{19}(41)$.

TORSION

R. Kane (Oxford): For a finite H-space X, if $H^*(\Omega X)$ has no p-torsion, then $H^*(X)$ has no p^2-torsion.

J. Lin (Princeton): For a finite H-space X,

1) $\Pi_3(X)$ has torsion at most of order 2 (also Kane)

2) for p odd and $v(j) = 1 + p + \ldots + p^j$,

if $\mathcal{P}^{p^j} H^{2r(j)}(X) \equiv 0$ mod decomposables for all j, then $H^*(\Omega X)$ has no p torsion.

(Note: E_8 mod 3 slips by.)

GENUS

Mislin-Zabrodsky: For a finite H-space X, Genus of X = Genus of Y iff $X \times S_X \simeq Y \times S_Y$ where $S_X = \Pi S^{n_i}$ where n_i runs over the distinct integers in the type of X.

Zabrodsky: The Genus of finite H-space X can be described as a quotient of a certain explicit finite group.

OTHER LOCAL RESULTS

G.H. Toomer (Cornell): Results on mod p category and co-category. There are some surprises.

Arkowitz, C.P. Murley & A. Shar (UNH): Results on the number of multiplications on an H-space in relation to the number on the localizations.

CLASSIFYING SPACES

There continue to be refinements of our understanding and application of classifying spaces

a) for local objects such as foliations (Haefliger-Bott-Shulman-Stasheff)

b) for specialized fibrations defined globally (May).

Hubbuck, Mahmud and Adams: Results on the classification of maps $BG \to BH$ for compact connected Lie groups in terms of Dynkin diagrams

Gitler and Feder: Necessary conditions for a map $HP(n) \circlearrowleft$ of degree d.

INFINITE LOOP SPACES

The exposition of infinite loop structures is reaching maturity. In addition to May: Geometry of Iterated Loop Spaces, there have recently appeared

Boardman-Vogt: Homotopy Invariance of Algebraic Structures and

May. E_∞-spaces,... LMS Lecture Notes 11.

There is also the approach of Segal which is particularly appealing for algebraic K-theory. Anderson's adaptation is further transmuted in Rector's talk. Segal's latest preprint version of "Categories and cohomology theories" implies a technical comparison between his Γ-spaces and the E_∞-spaces of Boardman-Vogt and May. Another approach to the comparison and to the homotopy invariance is due to

T. Lada (NCSU). A more general sythesis was initiated by Floyd at a regional conference at Binghamton (SUNY) in October, Proceedings to appear.

V. Snaith (Cambridge U) has fit Dyer-Lashof operations into K-theory.

SH-ALGEBRA

Gugenheim has shown that Sullivan's unit interval notion of homotopy of maps of DGA algebras does give a homotopy as shm-maps, assuming characteristic 0.

Halperin (Toronto) and Stasheff have constructed shm-maps $H^*(X) \to C^*(X)$ which are homotopy equivalences for any free commutative $H^*(X)$ of finite type.

Finally one of the biggest headlines this winter has been the Becker-Gottlieb proof of the Adams conjecture, notable not only for its simplicity but also for its strong use of the maximal torus and normalizer in $O(n)$. Reduction of the problem to the normalizer uses their generalization of transfer. The result on the normalizer is the equivariance of Adam's original proof for line bundles.

THE MOD p DECOMPOSITION OF LIE GROUPS

JAMES DILLON STASHEFF

TEMPLE UNIVERSITY

The theories of localization and completion in homotopy theory have given us important tools for pursueing the homotopy theory of Lie groups. One striking phenomenon is the mod p decomposability of simple Lie groups. To some extent, this was already apparent from the point of view of Serre's homotopy theory modulo classes of abelian groups. In particular, Hopf's result on rational cohomology can be interpreted as a rational equivalence between the Lie group G and a product of odd dimensional spheres:

$$G \underset{0}{\sim} \prod S^{2n_i-1}$$

The number of spheres is the rank of G.

For low rank, [Serre] showed the result held mod p. The next general result was due to [Harris]. He showed essentially that

$$U(2n) \underset{p}{\simeq} Sp(n) \times U(2n) / Sp(n)$$

for all odd primes p. For p = 3, the factors are indecomposable.

In a series of papers, [Mimura and Toda] exhibited the mod p decomposability of the torsion free Lie groups of moderate rank. [A rather different approach is used by Harper, Wilkerson and Zabrodsky to decompose any finite H-complex with cohomology of a reasonably restricted form]. The ultimate for torsion free Lie groups is provided by [Nishida] using an entirely different construction.

Theorem (Nishida). Given a prime p, for all r and n, there exist finite simply connected complexes $X_r(n)$ such that

$$U(n) \;{}_{p} \prod_{r=1}^{p-1} X_r(n)$$

$$S_p(n) \;{}_{p} \prod_{r=1}^{p-1/2} X_{2r}(n)$$

For an exceptional Lie group G without torsion, G is mod p equivalent to an appropriate product of $X_r(n)$'s. The spaces $X_r(n)$ are indecomposable except $X_1(n) = S^1 \times \overline{X}_1(n)$ with $\overline{X}_1(n)$ indecomposable.

U(n) plays a special role in this discussion, so I'll confine myself to that case. Let's see what the decomposition looks like in relation to the usual bundle

$$U(n-1) \to U(n) \to S^{2n-1}.$$

First of all $X_r(n) = X_r(n+1)$ for $k(p-1) < n < (k+1)(p-1)$. If we write $n = k(p-1)+s$ with $0 < r \leq p-1$, then the theorem asserts

$$U(n) \underset{p}{\sim} \prod_{s+1}^{p-1} X_r\ (k(p-1)) \times \prod_{1}^{s} X_r\ ((k+1)(p-1))$$

Notice that $X_r(p-1) = S^{2r-1}$.

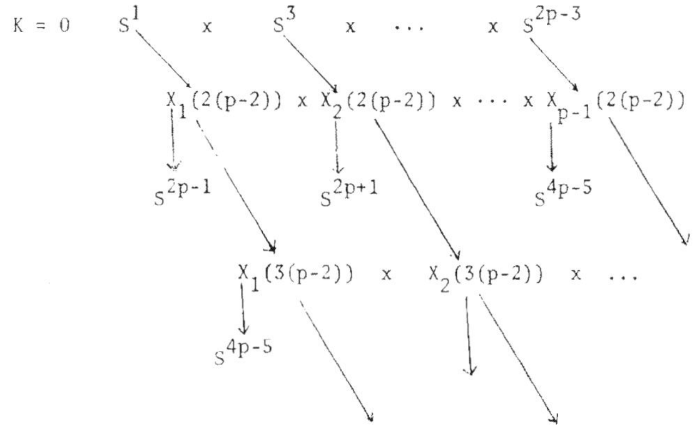

Here each piece $X_r((k-1)(p-1))$

$$\begin{array}{ccc} X_r((k-1)(p-1)) & \dashrightarrow & X_r(k(p-1)) \\ & & \downarrow \\ & & S^{2k(p-1)+2r-1} \end{array}$$

is a bundle [Nishida].

NISHIDA'S CONSTRUCTION

Some very elementary number theory plays an important role. A number q is a primitive root mod p if $q^{p-1} \equiv 1$ (p) but no smaller power is. It follows that $q^i - q^r \equiv 0(p)$ iff p-1 divides i-r. For a given n, let us fix such a $q > n$ until further notice.

For $q > n$, [Sullivan] has constructed maps

$$\psi^q: \ BU(n) \to BU(n) \text{ such that } \psi^q c_i = q^i c_i$$

where c_i is the i-th Chern class. Let $f_s: V(n) \to V(n)$ be devined as the p-localization of $\overline{f}_s(x) = \Omega\, \psi^q(x) \cdot x^{-q^s}$. Recall that with Z/p or Z_p - coefficients $H^*(V(n)) \cong E(h_1,\dots,h_n)$ where h_i trangresses to c_i. Hence $f_s^* h_i = (q^i - q^s)h_i$.

Now fix r such that $1 \leq r \leq p-1$ and consider the sequence $(f_1, f_2,\dots,\hat{f}_r,\dots, f_n, f_1, f_2,\dots)$:

$$V(n) \xrightarrow{f_1} V(n) \xrightarrow{f_2} \cdots V(n) \xrightarrow{f_{r-1}} V(n) \xrightarrow{f_{r+1}} V(n) \ldots \to V(n) \xrightarrow{f_n} V(n) \xrightarrow{f_1} V(n) \to \cdots$$

where f_i is consistently omitted for $i \equiv r(p-1)$. Let T(r,n) be the mapping telescope of this sequence. It is not hard to compute $H^*(T(r,n))$ in terms of $\varprojlim f_s^*(h_i)$. For $i \not\equiv r\ (p-1)$, the class h_i is repeatedly mapped to zero in the sequence,namely by f_i. For $i \equiv r$ mod p-1 and with Z/p or Z_p as coefficients, the class h_i is mapped isomorphically(multiplied by a unit) under each f_s in the sequence. Thus we have

$$H^*(T(r,n);Z/p) \cong E(h_r, h_{r+(p-1)},\dots,h_{r+k(p-1)})$$

where $k = [\frac{n}{p-1}]$. These cohomology statements imply T(r,n) is a finite p-local space since it is simply connected. [We are most grateful to Clarence Wilkerson for this alternative to certain ambiquities in Nishida's treatment at this point.] Now assemble $T(r,n)_p$ with $U(n)_\ell$ for the set of primes ℓ different from p to obtain $X_r(n)$ of the homotopy type of a finite complex. The initial inclusion of $U(n)$ into $T(r,n)$ combines with the identity to give a map $U(n) \to X_r(n)$. The product

$$U(n) \to \prod_{1}^{p-1} X_r(n)$$

is a mod p (and hence p-local equivalence) since the map in cohomology on each factor $X_r(n)$ is an isomorphism onto the sub-algebra generated by $\{h_{k(p-1)+r}\}$.

The proof of irreducibility uses the Hurewicz homomorphism to show there can not be more than p-1 factors in SU(n). The decomposition is unique in terms of homotopy type since [Wilkerson's] result applies.

SUB - FINITE - GROUP SPACES

Since we have decomposed U(n) at p as spaces, it is natural to ask if the factors are group spaces at p. [Ewing] has extensive negative results on this question and shows still more generally that the splitting is not as groups. In the positive direction certain factors or collections of factors are particularly interesting. Methods of Sullivan and Quillen have been applied by several people to show $T(p-1, n)$, that is $X_{p-1}(n)$ at p, is a loop space and I can show the map $T(p-1,n) \to V(n)$ is a loop map, in fact a sub p-finite group space (in the sense of Rector). For $p = 3$, this is the familiar subgroup inclusion $S_p([n/2]) \to U(n)$. [Clark, Ewing, and Wilkerson have used the method more extensively to construct several new "exceptional" groups].

To emphasize the comparison with Lie groups, replace the maximal torus BT^n by $\hat{BT}^n = K(\hat{Z}_p, 2)^n$ where $\hat{Z}_p$ denotes the p-adic integers. Let the Weyl group $W = \Sigma_n \wr Z/_{p-1}$ act in the obvious way: Σ_n permutes the factor and $Z/_{p-1}$ acts as an automorphism group independently on each factor $K(\hat{Z}_p,2)$. It is not hard to compute the invariant algebra $H^*(B\hat{T}^n)^W$ with Z/p coefficients. It is what we expect of $BT(p-1,n)$ namely

$$Z/p[c_{p-1}, c_{2(p-1)}, c_{2(p-1)},\ldots,c_{k(p-1)}],\ k = [\tfrac{n}{p-1}].$$

The desired classifying space then could be the homotopy quotient $Y(n)$ of $B\hat{T}^n$ (i.e. $EW \times_W B\hat{T}^n$ where EW is the universal principal W-bundle) except that it is not simply connected. For $n < p$, the p-completion $Y(n)^\wedge_p$ gives what we are after. For larger values of n, we turn to the work of Quillen on classifying spaces for finite fields.

Let F_q denote the field with q elements (so that $q = (q^1)^f$ for some prime q^1). Let d be the order of q mod p, i.e. the smallest integer such that $p | q^d - 1$.

Thm [Quillen]. If q is a prime different from p and d is the order of q mod p, then

$$H^*(BGL_{nd}(F_q); Z/p) \cong$$
$$Z/p[c_d, c_{2d},\ldots,c_{nd}] \otimes E(h_d, h_{2d},\ldots,h_{nd})$$

where $c_{jd} = \beta h_{jd}$ for some (possibly higher order) Bockstein β. (See [Quillen] for more precision).

The generators are called c_{jd} because they are the images of Chern classes pulled back according to a representation $GL_{nd}(F_q) \to GL_\infty(\mathbb{C})$.

Quillen proves this theorem by considering a Weyl-group situation. Let $k = F_q(\mu_p)$ denote the extension obtained by adjoining the p-th roots of unity. The degree of this extension is d. Let σ be a generator of the Galois group Z/d. It is known that $GL_n(k(\sigma)) \cong GL_{nd}(F_q)$. We regard $(k^*)^n \subset GL_{nd}(F_q)$ as diagonal blocks and consider the normalizer N. Clearly $N/(k^*)^n$ contains the wreath product $W = \Sigma_n \int Z/d$. Thus we have $(k^*)^n \to N \to N/k^{*n} \to (Bk^*)^n$ and hence $EW \times_W (Bk^*)^n \to BN \to BGL_{nd}(F_q)$ which composite Quillen shows maps Z/p-cohomology isomorphically to the W-invariant sub-algebra. Notice that $k^* = Z/_{q^d-1}$ so $Bk^* \simeq K(Z/_{q^d-1},1)$. For the mod p cohomology we are only interested in the p-primary part: $K(Z_p r,1)$. We have the induced action of σ and hence all of Z/d on $K(Z_p r,1)$.

Theorem [Quillen]:

(*) $H^*(K(k^*,1)^n/\Sigma_n \int Z/d; Z/p) \to H^*(K(k^*,1)^n; Z/p)^W \simeq Z/p\,[c_d,\ldots,c_{nd}] \otimes E[h_d,\ldots,h_{nd}]$ where $\beta h_{jd} = c_{jd}$ for the r-th order mod p Bockstein β.

To obtain a realization of the polynomial part alone, consider the sequence $F_q \to F_{q^p} \to F_{q^{p^2}} \to \ldots$ The extension $F_{q^p}(\mu_p)$ is still of degree d over F_{q^p} but where before $p^r | q^d-1$ but p^{r+1} did not, now we have $p^{r+1} | q^d-1$ but p^{r+2} does not. Thus in (*) the form remains the same but the classes c_{jd} are in fact mapped isomorphically (consider the virtual factorization $F_q \to F_{q^p} \to \mathbb{C}$) while the classes h_{jd} are mapped to zero and replaced by new ones. We denote the limit field by $\bar{F}_q$; the space we seek is almost $BGL_{nd}(\bar{F}_q)$ since the cohomology is $Z/p[c_d,\ldots,c_{nd}]$; only the fundamental group $GL_{nd}(F_q)$ is in the way. This group however is p-perfect, i.e. its abelianization has no p-primary part, so the p-completion $BGL_{nd}(F_q)^{\hat{}}_p$ is a simply connected p-complete space with the desired cohomology. The pullback according to Sullivan's homotopy adele square gives the p-local space we are after: call it $BGL_{nd}(F_q)^*$.

The comparison with the Sullivan-type construction is easy; one is the Bockstein of the other. That is, the Bockstein of the sequence

$$\hat{Z}_p \to \hat{Z}_p \to k^* \simeq Z/q^{p-1}-1$$

induces $Bk^* \to K(\hat{Z}_p,2)$ and hence $(Bk^*)^n/W \to K(\hat{Z}_p,2)^n/W$.

As q gets large through the sequence q^{p^r}, this converges to a Z/p-cohomology isomorphism.

The comparison with $BU(n)$ is more subtle. Quillen's representation $GL_n(F_q) \to GL_\infty(\mathbb{C})$ needs to be factored at least up to homotopy through $GL_n(\mathbb{C}) \simeq U(n)$. To effect this, consider the algebraic closure Λ of Z_p in $\hat{\mathbb{C}}$ which has F_q as residue field. The induced maps $BGL_n(F_q) \leftarrow BGL_n(\Lambda) \to BGL_n(\mathbb{C})$ can be studied via étale homotopy theory and according to Artin-Mazur, Lubkin induce isormorphisms of Z/n cohomology for $(n,q) = 1$, in particular for Z/p.

This still holds after p-completion and adelization. In particular for $d = p-1$ and $m = n(p-1)$, the composite $\Omega BGL_m(F_q)^* \to U(m) \to X_{p-1}(m)$ is clearly a p-equivalence (just compute using the Serre or Eilenberg-Moore spectral sequence), so at p we have realized $X_{p-1}(m)$ as a sub-p-finite group space of $U(m)$. The other divisors d of p-1 give a lattice of sub-p-finite group spaces, barely hinted at by the classical

$$Sp(n) \to U(2n) \to O(4n).$$

These p-finite group spaces $X_d(n)$ clearly play a central role in the p-theory of group spaces; whether they generate the whole story remains to be seen.

BIBLIOGRAPHY

J. Ewing, The non-splitting of Lie groups as Loop spaces, preprint.

B. Harris, On the homotopy groups of classical groups, Ann. of Math 74 (1961)

M. Mimura and H. Toda, Cohomology operations and the homotopy of compact Lie groups I, Topology 9 (1970).

G. Nishida, On a result of Sullivan and the mod p decomposition of Lie groups, Research Inst. for Math. Sci, Kyoto U, (1971) (Mimeo).

D. Quillen, The K-theory associated to a finite field, I. (preprint)

J.-P. Serre, Groupes d'homotopie et classes de groupes abéliens, Ann. of Math 58 (1953).

D. Sullivan, Geometric Topology I, Notes, MIT (1970).

SELF-MAPS OF CLASSIFYING SPACES

Clarence Wilkerson
Carleton University
Ottawa, Ontario

The purpose of this note is to extend the Sullivan construction [19] of "unstable" Adams operations on $BU(n)$ to cover all compact connected semi-simple Lie groups, including the exceptional simple Lie groups. The application in this paper is to obtain mod p splittings of the simple Lie groups, but the result also provides a complement to the work of Hubbuck [8] studying the nonexistence of certain self-maps of classifying spaces.

Theorem I: If G is a compact connected semi-simple Lie group, there exists an "unstable" Adams operation $\psi^p: BG_{P-p} \to BG_{P-p}$ *with the property that* $\psi^{p^*} \mid H^{2n}(BG_{P-p}Q) = p^n \cdot Id$. *Here* BG_{P-p} *denotes the localization of BG away from the prime p.*

Corollary II: If W(G) is the Weyl group of G and p does not divide the order of W(G), then there is $\psi^p : BG \to BG$ *with the above property.*

A modification of the argument of Nishida [15,18] together with I. shows

Corollary III: Let G be a compact connected simple Lie group such that $H_*(G,Z)$ *has no p-torsion. Then G is p-equivalent to a product of H-spaces* $X_i(G_p)$ *where each* $X_i(G)$ *is indecomposable mod p and the type* $\{2i_1-1,..2i_j-1\}$ *of* $X_i(G)$ *has the property that* $i_1=\ldots=i_j= i \bmod (p-1)$.

Theorem I has acquired the status of a folk theorem, but in view of the applications, deserves recording. The approach here follows Sullivan [19] closely after the appropriate analogues of the Grassmanians are produced. The construction by Rector [16] of a $B^{et}G$ offers an alternative proof which is better suited in general for constructing exotic representations $BG \to BH$, but we choose the direct route for the purposes of this note.

Section I contains the proof of Theorem I and its corollary, while Section II is devoted to the applications. This work originated as a response to the talks of J. Stasheff and D. Rector at this conference. The author is happy to acknowledge helpful discussions also with J.Stasheff, D. Rector, G. Mislin, and J. Morava about this material.

Section One: Sullivan's Construction

We briefly review some terminology from algebraic geometry. A prescheme $X_{\mathbb{C}}$ over Spec $\mathbb{C}$ is *defined over* $\mathbb{Q}$ if $X_{\mathbb{C}}$ has the form $X'_{\mathbb{Q}} \times_{\mathbb{Q}}$ Spec $\mathbb{C}$, for some $X'_{\mathbb{Q}}$. A morphism of preschemes $f: X \to Y$ is *defined over* $\mathbb{Q}$ if X and Y are defined over $\mathbb{Q}$, and f is of the form $f'_{\mathbb{Q}} \underset{\mathrm{Id}_{\mathbb{Q}}}{\times} \mathrm{Id}_{\mathbb{C}}$ for some $f'_{\mathbb{Q}}: X'_{\mathbb{Q}} \to Y'_{\mathbb{Q}}$. If $X_{\mathbb{C}}$ is defined over $\mathbb{Q}$, then any $\sigma \in \mathrm{Gal}(\tilde{\mathbb{Q}}/\mathbb{Q})$ induces an automorphism of $X_{\mathbb{C}}$, $\sigma_* = \mathrm{Id} \underset{\mathrm{Id}}{\times} \tilde{\sigma}$, where $\tilde{\sigma}$ is an extension of σ to $\mathbb{C}$. If $f: X_{\mathbb{C}} \to Y_{\mathbb{C}}$ is defined over $\mathbb{Q}$, then $f\sigma_* = \sigma_* f$.

Theorem 1.1 (Artin-Mazur [1,2], Sullivan [19]): Let V be a prescheme over $\mathbb{C}$ *and* $V(\mathbb{C})$ *= its complex valued points =* $\mathrm{Hom}_{p\text{-}s}(\mathrm{Spec}\ \mathbb{C}, V)$ *with the strong topology. If* $V(\mathbb{C})$ *is a connected manifold, then* $V(\mathbb{C})\hat{} = (EH(V))\hat{}$. *Here* $\hat{}$ *is profinite completion with respect to all primes, and EH is the etale homotopy functor of Artin-Mazur collapsed to a homotopy type via Sullivan.*

Corollary 1.2 (Sullivan, Lubkin [12]):

i) If V is defined over $\mathbb{Q}$, *then* $\mathrm{Gal}(\tilde{\mathbb{Q}}/\mathbb{Q})$ *acts on* $V(\mathbb{C})\hat{}$.

ii) There is a homomorphism $\pi: \mathrm{Gal}(\tilde{\mathbb{Q}}/\mathbb{Q}) \to \hat{Z}^*$, *the units of the profinite completion of the integers,* $\hat{Z}$. π *is onto, with kernel the commutator subgroup.*

iii) If $\sigma \in \mathrm{Gal}(\tilde{\mathbb{Q}}/\mathbb{Q})$ *and* $\pi(\sigma) = \bar{a} \in \Pi \hat{Z}_p = \hat{Z}$, *then* $(\sigma_*)^* \mid H^{2n}(P^N(\mathbb{C})\hat{}, \hat{Z}) = \bar{a}^n \cdot \mathrm{Id}$.

The remainder of Sullivan's proof is apply Corollary 1.2 to the Grassmanians , since the direct limit of these give the classifying space of U(n). We wish to define now the proper analogue of the Grassmanians for any connected compact semi-simple Lie group G. There is a complex Lie group $G(\mathbb{C})$ for which G is a maximal compact subgroup, see for example Serre [17]. In fact, G is a strong deformation retract of $G(\mathbb{C})$ as groups, so BG and $BG(\mathbb{C})$ are homotopy equivalent. By Konstant [10], there is an algebraic group $G_{\mathbb{C}}$ defined over Z (hence $\mathbb{Q}$), for which $G(\mathbb{C})$ is the set of complex valued points with the strong topology. By Borel [3], $G_{\mathbb{C}}$ has a faithful representation $\zeta: G_{\mathbb{C}} \to GL(N,C)$ for some N, and ζ is defined over $\mathbb{Q}$. Consider $GL(N,\mathbb{C})$ to be embedded in $GL(N+n,\mathbb{C})$ in the upper left hand corner, and $GL(n,C)$ in the lower right hand corner. Then $GL(N+n,\mathbb{C}) \to GL(N+n,\mathbb{C})/I \times GL(n,\mathbb{C}) \to GL(N+n,\mathbb{C})/\zeta G_{\mathbb{C}} \times GL(n,C)$ are all defined over $\mathbb{Q}$. The last is principal $G(\mathbb{C})$-bundle when we take the complex valued points. Then $GL(N+n,\mathbb{C})/GL(n,\mathbb{C})$ is a complex Stiefel manifold and its connectivity increases as n increases. This completes the proof of the next theorem.

Theorem 1.3: If G is a compact connected semi-simple Lie group, then $BG = \varinjlim GL(N+n,\mathbb{C})/\zeta G(\mathbb{C}) \times GL(n,\mathbb{C})$ *and hence* $\mathrm{Gal}(\tilde{\mathbb{Q}}/\mathbb{Q})$ *acts on* $BG\hat{}$.

Corollary 1.4: If $\sigma \in Gal(\bar{Q}/Q)$ *with* $\pi(\sigma) = \bar{a}$, *then* $(\sigma_*)^*|H^{2n}(BG\hat{}, \hat{Z} \otimes Q) = \bar{a}^n \cdot Id.$

Proof: G_C has a maximal torus T defined over Q, and $H^*(BC\hat{}) \to H^*(BT\hat{})$ is monic for these coefficients. Since the map is also equivariant with respect to the Galois action, this determines $(\sigma_*)^*$.

Corollary 1.5: There exists ψ^p: $BG_{P-p} \to BG_{P-p}$ *with the desired property.*

Proof: There is a $\sigma_p \in Gal(\bar{Q}/Q)$ with $\pi(\sigma_p) = \bar{a} = (p,..1,p...)$ in $\hat{Z} = \Pi \hat{Z}_q$. That is, all coordinates are p, except the p-th coordinate, which is 1. By definition , $\bar{a} \in \hat{Z}^{*}$. The arithmetic square below is a fibre square up to homotopy, since the homotopy groups of BG are finitely generated; Sullivan [19] and Bousfield-Kan [5].

$$\begin{array}{ccc} BG_{P-p} & \xrightarrow{r_p} & BG\hat{}_{P-p} \\ r_0 \downarrow & & \downarrow i_p \\ BG_0 & \xrightarrow{i_0} & (BG\hat{}_{P-p})_0 = \Pi K(2r_i, (\Pi_{q \neq p} \hat{Z}_q) \otimes Q) \end{array}$$

Hence maps f_0: $BG_{P-p} \to BG_0$ and f_p: $BG_{P-p} \to BG\hat{}_{P-p}$ with $i_0 f_0 \simeq i_p f_p$ determine (not uniquely , since BG is not a finite complex) a map f: $BG_{P-p} \to BG_{P-p}$. Taking $f_p = \sigma_{p*} r_p$ and $f_0 = r_0$ followed by the p-th power map on each factor of $BG_0 = \Pi K(2r_i, Q)$, we see that this condition is satisfied and denote f as ψ^p .

Corollary 1.6: If $p \nmid |W(G)|$, *then there exists* ψ^p: $BG \to BG$ *with the desired property.*

Proof: In view of the fibre square (up to homotopy)

$$\begin{array}{ccc} BG & \to & BG_{P-p} \\ \downarrow & & \downarrow \\ BG_p & \to & BG_0 \end{array}$$

it suffices to find a ψ^p: $BG_p \to BG_p$ with $\psi^{p*}|H^{2n}(BG_p, Q) = p^n \cdot Id.$ By Sullivan, Mislin, and Wilkerson, $BG_p = (BT/W(G))_p$ and the p-th power map on $K(Z^r, 2)$ induces a ψ^p: $BG_p \to BG_p$ with the right property.

The ψ^p we have constructed is not a priori unique, since choices were made in the selection of σ_p and in the liftings in the two fibre squares. However, in the simplest case $BSU(2) = BS^3$, it is unresolved whether such a map is unique up to homotopy.

Section Two: Applications

Theorem 2.1 is a generalization of Nishida [15], which modified the argument giving the mod p splitting of BU in terms of the eigenvalues of the Adams operations. It is also an illustration of a phenomena observed in Wilkerson [21]; namely, that if a finite H-space has a self-map which does not induce a multiple of the identity or the zero map on cohomology when iterated, then the H-space is mod p decomposable . The given map gives a practical means of computing the decomposition, in fact. After the proof, we use 2.1 to give the best possible mod p splittings of the exceptional Lie groups at the primes p for which the homology has no torsion . For E_8 mod 5 it also works and gives a previously unknown decomposition. This proceeds in fairly painless manner compared to the hard homotopy calculations previously required, see [13,14], [22].

Theorem 2.1: Let Y be a finite H-space and $\Phi : Y \to Y$ such that $\Phi^* \mid QH^{2n-1}(Y,Q) = q^n \cdot Id$ for all $n > 0$. If q is a primitive $(p-1)$-st root of unity mod p, and $H_*(Y,Z)$ has no p-torsion, then $Y_p = \Pi X_i(Y_p)$ where the type $\{2i_1-1, \ldots 2i_j-1\}$ of $X_i(Y_p)$ has the property that $i_1 = \ldots = i_j = i$ mod$(p-1)$ and the product is taken over all residue classes mod$(p-1)$.

<u>Proof</u>: Let $0 < i \leq p-1$ be given and $\{2r_j-1\}$ be the entries of the type of Y such that $r_j \not\equiv i$ mod$(p-1)$. Define $f_{r_j} = \Phi - \mathrm{Pow}(q^{r_j})$, where $\mathrm{Pow}(q^{r_j})$ is the q^{r_j}-st power map on Y and the addition is defined using the H-space structure of Y. Then $f^{r_j *} \mid QH^{2n-1}(Y_p, Z_p) = q^n - q^{r_j}$. Thus $f^{r_j *} \mid QH^{2r_j-1}(Y_p, Z_p) = 0$ and $(f_{r_j})_* = 0$ on $\pi_{2r_j-1}(Y_p)/$ torsion. $f^{r_j *}$ is an isomorphism on indecomposables in dimension $2n-1$ if and only if $n \not\equiv r_j$ mod$(p-1)$. Finally, define F_i to be the composite in any order of the f_{r_j} for which $r_j \not\equiv i$ mod$(p-1)$, and let $X_i(Y_p)$ be the infinite mapping telescope of $Y_p \xrightarrow{F_i} Y_p \xrightarrow{F_i} Y_p \to \ldots \to$. Then $\pi_n(X_i(Y_p)) = \varinjlim (\pi_n(Y_p), F_{i*})$ is a finitely generated Z_p-module for all n. (Nishida's original argument seems to produce Q's for some of the infinite homotopy groups.). The condition on the types is clearly true if the $X_i(Y_p)$ are H-spaces. But there is a map $Y_p \to X_i(Y_p)$ and hence to the product. It is easy to see that this induces an isomorphism in $H^*(Z/p)$. Since Y_p and $X_i(Y_p)$ have homotopy groups finitely generated over Z_p, the map $Y_p \to \Pi X_i(Y_p)$ is a homotopy equivalence.

For a Lie group G, we can take $\Phi = \Omega\psi^q$ by theorem I. We record the decompsitions of the simple exceptional Lie groups obtained by applying theorem 2.1. The solid arrows indicate nontrivial P^1's and the dotted arrows are secondary operations described in Prop.2.3 ([11], [13]):

G_2: type (3, 11)
$p = 3$: (3 --→ 11)
$p = 5$: (3 → 11)
$p \geq 7$: (3),(11)

F_4: type (3, 11, 15, 23) : 2 and 3-torsion
$p = 5$: (3 → 11), (15 → 23)
$p = 7$: (3 → 15), (11 → 23)
$p = 11$: (3 → 23), (11),(23)
$p \geq 13$: (3),(11), (15),(23)

E_6: type (3, 9, 11, 15, 17, 23) : 2 and 3-torsion
$p = 5$: (3 → 11), (9 → 17), (15 → 23)
$p = 7$: (3 → 15), (9),(11 → 23), (17)
$p = 11$: (3 → 23), (9),(11),(15), (17)
$p \geq 13$: (3),(9), (11),(15),(17), (23)

E_7: type (3, 11, 15, 19, 23, 27, 35) : 2 and 3-torsion
$p = 5$: (3 → 11 --→ 19 → 27 → 35), (15 → 23)
$p = 7$: (3 → 15 --→ 27),(11 → 23 → 35),(19)
$p = 11$: (3 → 23),(11),(15 → 35),(19),(27)
$p = 13$: (3 → 27),(11 → 35),(15),(19),(23)
$p = 17$: (3 → 35),(11),(15),(19),(23),(27)
$p \geq 19$: (3),(11),(15),(19),(23),(27),(35)

E_8: type (3, 15, 23, 27, 35, 39, 47, 59) : 2,3, and 5-torsion
$p = 7$: (3 → 15 --→ 27 → 39), (23 → 35 → 47 → 59)
$p = 11$: (3 → 23),(15 → 35), (27 → 47),(39 → 59)
$p = 13$: (3 → 27),(15 → 39), (23 → 47),(35 → 59)
etc.
$p \geq 31$: (3),(15),(23),(27), (35),(39),(47),(59)

Theorem 2.1 can also be applied to the irreducible symmetric spaces, such as E_6/F_4, which Harris [7] has shown to be mod p odd H-spaces.

Corollary 2.2: [13] *If G is a compact simply-connected exceptional simple Lie group such that $4p-1$ does not occur in the type of G, then G is p-equival-*

ent to a product of spheres and $B_n(p)$'s. That is, p is quasi-regular for G. We assume here that $H_*(G,Z)$ has no p-torsion.

New Proof: From the table above, for these primes G has only rank one factors and rank two factors for which the generators of Z/p-cohomology are linked by a nontrivial P^1. The rank one factors are p-equivalent to spheres, and by Zabrodsky [23],[22] the rank two factors are p-equivalent to $B_n(p)$'s.

Proposition 2.3: If G is a compact simply-connected simple Lie group with $H_*(X,Z)$ p-torsion free, then $X_i(G_p)$ is indecomposable for all i.

Proof: This is proved by Nishida [15] for the classical simple groups. The P^1 operations cover all the exceptional groups except (G_2; p = 3), (E_7; p = 5, X_2: 11 ? 19), (E_7; p = 7, X_2: 15 ? 27), (E_8; p = 7, X_2: 15 ? 27). A modification of the counting arguments in Wilkerson [20] shows that there are nontrivial operations linking these generators. For dimensional reasons, it is a secondary operation associated to the vanishing of P^2x_3.

Proposition 2.4: a) $X_2((E_7)_5) = X_2(SU(18)_5)$

b) $X_2((E_7)_7) = X_2(SU(14)_7)$

c) $X_5((E_7)_7) = X_5(SU(18)_7)$

d) $X_2((E_8)_7) = X_2(SU(20)_7)$

Proof: There are at least three approaches to the proof. The first is not sporting: since the factors are indecomposable, they must agree with those found by Mimura-Toda [14], by the unique factorization theorem of [21]. In the second method, we adopt the approach of [14], but only for identifying the factors, not to achieve the factorization. Namely, we have the adjoint representations $f_j: E_j \to SU(N)$. By tedious computations, one can calculate the induced map in cohomology in terms of the roots of E_j. $H^*(f_j, Z/p)x_m \neq 0$ for (E_7, p=5 or 7, m=3, p=5, m=11) and for (E_8, p=7, m=3). In view of the cohomology operations, this implies that the induced maps g_j: $X_i(E_{j,p})$ induce cohomology isomorphisms through the dimensions of the primitives of $X_i(E_{j,p})$. The inclusions $SU(n) \to SU(N)$ have the same properties. By the lifting theorem of Zabrodsky [23, 22], we can lift g_j through the relative Postnikov tower of $X_i(SU(n)_p) \to X_i(SU(N)_p)$ to a p-equivalence in all the above cases except E_7, p = 5, i = 2. There are five generators, so in this case there is a possible obstruction in dimension 95. The third method follows from the ideas of Zabrodsky [24]. Using the Ω-spectrum for BP-cohomology, we can construct models of these H-spaces. If $BP(2n+1,p)$ is the 2n-connected stage in the spectrum, then $H^*(BP(2n+1,p),Z_p) = \Lambda(x_{2n+1}, x_{2n+1+d},\ldots)$, where $d = 2(p-1)$, up to dimension $2n+1+pd$. We have H-maps of $X_2(G_p) \to BP(3,p)$ which classify $BP^3(X_2(G_p))$ generators. From the cohomology operations, these induce Z/p-equivalences through the dimensions of the primitives of $X_2(G_p)$. The previous argument then applies to lifting from $X_2(E_{j,p})$ to $X_2(SU(n)_p)$. The same techniques work for $X_5(E_7)_7)$ and to show that $X_5(E_8)_7$ is the unique H-space with its cohomology (including operations). These are discussed in more detail in [22]. These calculations are based on Steve Wilson's results on BP.

If G has torsion, Theorem 2.1 can still be applied if it can be verified that $H^*(G,Z/p) = \otimes H^*(X_i(G_p),Z/p)$. The exceptional groups amenable to this are $(E_6,\ p = 3)$ and $(E_8,\ p = 5)$.

Proposition 2. 3: a) $H^*(X_1((E_6)_3),Z/3) = \Lambda(x_9,\ x_{17})$

$H^*(X_2((E_6)_3),Z/3) = \Lambda(x_3,x_7,x_{11},x_{15}) \otimes Z/3[x_8]/[x_8^3]$

with $P^1x_3 = x_7,\ \beta_3x_7 = x_8$.

b) $H^*(X_0((E_8)_5),Z/5) = \Lambda(x_{15},x_{23},x_{39},x_{47})$

$H^*(X_2((E_8)_5),Z/5) = \Lambda(x_3,x_{11},x_{27},x_{35}) \otimes Z/5[x_{12}]/[x_{12}^5]$.

with $P^1x_3 = x_{11},\ \beta_5x_{11} = x_{12}$.

Proof: Use Borel's calculations [4] of these cohomology rings and the self-maps Γ_i defined in 2.1.

This decomposition of E_8 mod 5 is new. It seems likely that X_0 is indecomposable and that X_2 splits off a further factor, namely $B_{19}(5)$. Unfortunately, a) was previously known. By Harris [7], $E_6 =_p F_4 \times E_6/F_4$ for odd p. By Conlon [6], $E_6/F_4 = X_1((E_6)_3)$ is indecomposable since $\pi_{16}(E_6/F_4) = 0$. Harper [9] has found a mod 3 decomposition of F_4 with one factor being $B_5(3)$. The exceptional groups E_7 and E_8 mod 3 remain unexplored at present. I would like to thank John Harper for pointing out the work of Harris [7].

References:

1) Artin, M.: SGA IV(1963-4), expose XI, Lecture Notes in Mathematics, vol. 305, Springer (1970).
2) Artin, M., Mazur, B.: Etale Homotopy, Lecture Notes in Mathematics, vol. 100, Springer (1969).
3) Borel, A.: Linear Algebraic Groups, Benjamin (1969).
4) Borel, A.: Sous-groupes commutatifs et torsion des groupes compact connexes, Tohoku J. Math. 13(1961), 216-240.
5) Bousfield, A., Kan, D.: Homotopy Limits, Completions, and Localizations, Lecture Notes in Mathematics, vol. 304, Springer (1972).
6) Conlon, L.: An application of the Bott suspension map to the topology of EIV, Pacific J. Math. 19(1966), 411-428.
7) Harris, B.: Suspensions and characteristic maps for symmetric spaces, Ann. of Math. 2(76)(1962), 295-305.
8) Hubbuck, J.R.: S.H.M. self-maps of the classical simple Lie groups, preprint.
9) Harper, J.: A mod 3 splitting of F_4, this conference.
10) Konstant, B.: Groups over Z, in Algebraic Groups and Discontinuous Subgroups, PSPUM IX (1966) A.M.S., 90-98.
11) Kumpel, P.G., Jr.: Lie groups and products of spheres, P.A.M.S. 16(1965) 1350-6.
12) Lubkin, S.: On a conjecture of Weil, Amer. J. Math. 89(1967), 443-548.
13) Mimura, M., Toda, H.: Cohomology operations and the homotopy of compact Lie groups, Topology 9(1970), 317-336.
14) Mimura, M., Toda, H.: COHCLG-II, preprint.
15) Nishida, G.: On a construction of Sullivan, preprint.
16) Rector, D.: The Adams conjecture for spectra, this conference.
17) Serre, J-P.: Algebres de Lie semi-simples complexes, Benjamin (1966).
18) Stasheff, J.: Exotic mod p loop spaces: Nishida vs. Quillen, this conference.
19) Sullivan, D.: Localization, Periodicity, and Galois Symmetry, Geometric Topology I Notes, MIT 1970, revised 1971.
20) Wilkerson, C.: K-theory operations in mod p loop spaces, Math. Z. 132(1973), 29-44.
21) Wilkerson, C.: Genus and cancellation for H-spaces, this conference.
22) Wilkerson, C., Zabrodsky, A.: Quasi-regular primes for mod p H-spaces, preprint.
23) Zabrodsky, A.: On mod p odd rank two H-spaces, to appear.
24) Zabrodsky, A.: BP Adams resolutions for torsion-free H-spaces, in preparation.

GENUS AND CANCELLATION FOR H-SPACES

by

Clarence Wilkerson

Carleton University

Ottawa, Ontario

In Hilton-Roitberg [3], an H-space E_{7w} was constructed with the properties that $E_{7w} \times S^3 \simeq Sp(2) \times S^3$ and $E_{7w} \times E_{7w} \simeq Sp(2) \times Sp(2)$, but yet $E_{7w} \not\simeq Sp(2)$. That is, cancellation does not hold for the semi-group operation of cartesian product on finite H-spaces, even up to homotopy type. It was clear, however, that E_{7w} and $Sp(2)$ had homotopy equivalent p-localizations for all primes p. This prompted the definition by Mislin [4] that two nilpotent CW-complexes X and Y are in the same genus $(G(X) = G(Y))$ if and only if their p-localizations, X_p and Y_p, are homotopy equivalent for all primes p. It was conjectured by Mislin [4] and Hilton that the relation between genus and non-cancellation is always as illustrated by the E_{7w} example. Zabrodsky [6] has recently shown that if X is a finite H-space such that its genus G(X) contains more than one homotopy type, non-cancellation examples occur. The result announced here is a converse. A generalization and full details appear elsewhere. In the following, p is a fixed prime number.

Theorem I: 1) <u>If X is a 1-connected H-space with finitely generated homology, then $X_p \simeq \prod_{i=1}^{t} X_i$, where the X_i are irreducible p-local H-spaces and are unique up to order.</u>

2) <u>For X, Y, and W as in 1), if $X \times W \simeq Y \times W$, then</u> $G(X) = G(Y)$. <u>If there exists</u> $k > 0$ <u>with</u> $X^k = Y^k$, <u>then</u> $G(X) = G(Y)$.

Here X_i is <u>irreducible</u> if it has no nontrivial retracts. Since X_i is an H-space, this is equivalent to requiring that X_i have no nontrivial cartesian product factors, i.e. to X_i being <u>indecomposable</u>. In view of the finiteness assumptions on X, it automatically factors into a finite product of indecomposable H-spaces, and hence the important content of Theorem I(1) is the uniqueness of this factorization. I(2) is a consequence of this uniqueness. The results can also be stated more categorically as

Theorem II. Let lcpfdH be the homotopy category of 1-connected H-spaces with reduced integral homology finitely generated over the p-local integers $\mathbb{Z}_p$. Then the cartesian product operation on lcpfdH yields a free semi-group generated by the irreducible spaces of lcpfdH.

There are valid dual assertions for finite co-H-spaces and the wedge product which generalize the results of Freyd [1], [2] for the stable homotopy category.

I would like to express my appreciation to Guido Mislin and Alex Zabrodsky for many tutorials and also for access to their current research. In particular, this work was inspired by the partial solution of Mislin [5], while Alex Zabrodsky supplied patches for the gaps and inaccuracies of my original proof.

SKETCH OF PROOF

We work within the homotopy categories lcpft and lcpfd of p-local CW-complexes with $H_*(\mathbb{Z}_p)$ of finite type and finite dimension over $\mathbb{Z}_p$ respectively. In lcpft, $f: X \to Y$ is a homotopy equivalence $\Leftrightarrow$ $H_*(f, \mathbb{Z}/p)$ is an isomorphism $\Leftrightarrow$ $H_*(f, \mathbb{Z}_p)$ is an isomorphism.

Definition 1. X lcpft is H*-prime if, for all $f: X \to X$, either f is a homotopy equivalence or, for all $n > 0$, there exists $N_n > 0$ such that

$$(\tilde{H}^*(f, \mathbb{Z}/p))^{N_n} = 0 \quad \text{for all} \quad * < n.$$

The term "prime" is justified by the analogy, provided by the next lemma, with the characterization of the prime natural numbers by the property that $p | ab \Rightarrow p | a$ or $p | b$.

Lemma 2. If $A, B \in$ lcpft and $X \to A \times B \to X$ is a retraction, then X is a retract of either A or B via the composite maps, provided that X is H*-prime.

Theorem 3. If $W \in$ lcpft admits factorizations $W \simeq \Pi W_i$ with each W_i H*-prime and also $W \simeq \Pi W'_j$ with each W'_j irreducible, then the factorizations $\{W_i\}$ and

$\{W'_j\}$ agree up to order (and homotopy equivalence).

To prove Theorem I(1) it suffices then to show that irreducible H-spaces are H*-prime, i.e. that non-H*-prime H-spaces have nontrivial retracts. We need a non-tautological recognition principle for retracts in lcpfd:

Definition 4. Let $X \in$ lcpfd. Then $f: X \to X$ is a pseudo-projection $\Leftrightarrow$ image $H^*(f, \mathbb{Z}_p) = \text{image } (H^*(f, \mathbb{Z}_p))^2$.

Theorem 5. If $f: X \to X$ is a pseudo-projection with $\ker \tilde{H}^*(f, \mathbb{Z}/p) \neq 0$ and $\neq \tilde{H}^*(X, \mathbb{Z}/p)$, then X has a non-trivial retract.

The proof is a step-by-step construction of a Postnikov type approximation to the desired retract. Finally,

Theorem 6. If X is a non-H*-prime H-space in lcpfd, there is a non-trivial pseudo-projection $g: X \to X$.

Idea. Since X is a finite H-space, there is $p^r > 0$ such that for any endomorphism A of $QH^*(X, \mathbb{Z}_p)$/Torsion, there is $h_A: X \to X$ with $h_A^*|QH^*/\text{Torsion} = p^r A$ and $\tilde{H}^*(h_A, \mathbb{Z}/p) = 0$. If $f: X \to X$ is not a homotopy equivalence or $\tilde{H}^*(f, \mathbb{Z}/p)$ nilpotent, then some iterate of f, f^M looks like a pseudo-projection on QH^*/Torsion mod p^r. The "deviation" $p^r A$ can be subtracted off by setting $f' = \mu(f^M, h_{-A})$, where μ is the H-space structure map. Some iterate of f' will be a pseudo-projection. Similar constructions apply to co-H-spaces and rational H-spaces.

REFERENCES

1. P. Freyd, Stable homotopy, Proceedings of the Conference on Categorical Algebra (La Jolla, 1965), Springer Verlag (1966).

2. P. Freyd, Stable homotopy II, AMS, PSPM Vol. XVIII (1970), pp. 161-183.

3. P. J. Hilton, J. Roitberg, On principal S^3-bundles over spheres, Ann. of Math. 90 (1969), 91-107.

4. G. Mislin, The genus of an H-space, Lecture Notes in Mathematics 249, Springer Verlag (1971), 75-83.

5. G. Mislin, Cancellation properties of H-spaces, to appear in Comm. Math. Helv.

6. A Zabrodsky, On the genus of finite CW H-spaces, to appear.

p EQUIVALENCES AND HOMOTOPY TYPE

Alexander Zabrodsky*
Institute for Advanced Study, Princeton, New Jersey

This study represents a generalization, extension and to some degree a simplification of the main theorem in Zabrodsky [11]. It answers some questions raised in Mislin [8] for a family richer than finite CW - H spaces.

All spaces considered are pointed and of the homotopy type of simply connected CW-complexes of finite type.

For a CW complex X let $G(X)$ be the set of homotopy types of spaces Y with $Y_p \approx X_p$ for every prime p (where $(\ )_p$ is the p-localization operation). The main result of this study is given by:

Main Theorem: Let X be an H_0 space, i.e.: $H^*(X,Q)$ is a free algebra. Suppose $\pi_n(X) = 0$ for $n \geq N(X)$.

There exists an integer t, depending on $\pi_*(X)$, $H_*(X,Z)$ and the Hurewicz homomorphism $\pi_*(X) \longrightarrow H_*(X,Z)$ so that the following holds: Let $[X,X]_t$ be the set of homotopy classes of maps $f : X \longrightarrow X$ so that $H^*(f,Z) \otimes Z_t$ is an isomorphism ($Z_t = Z/tZ$). Let $\alpha : [X,X]_t \longrightarrow [(Z_t^*)/\pm 1]^\ell$ be given by the composition

$$[X,X]_t \longrightarrow \mathrm{Aut}(QH^*(X,Z)/\mathrm{torsion} \otimes Z_t) \xrightarrow{|\det|} [(Z_t^*)/\pm 1]^\ell$$

where Z_t^* are the units in Z_t and ℓ is the number of integers n with $QH^n(X,Z)/\mathrm{torsion} \neq 0$. Then $G(X)$ admits an abelian group structure and one has the following exact sequence:

$$[X,X]_t \xrightarrow{\alpha} (Z_t^*/\pm 1)^\ell \xrightarrow{\xi} G(X) \longrightarrow 0.$$

In section 1 general properties of H_0 spaces are studied. The main theorem is proved in section 2 and in section 3 the main theorem is applied to answer some questions raised in Mislin [8], page 83, and to illustrate some computations of $G(X)$, $X = Sp(2)$, G_2.

*The author was partially supported by NSF Grant GP-36418X1.

1. H_0 SPACES AND p-EQUIVALENCES

1.1. <u>Definition</u>: An H_0 space is a CW complex X such that $H^*(X,Q)$ is a free algebra on generators $x_{n_1}, x_{n_2}, \ldots, x_{n_r}$, $\dim x_{n_i} = n_i, n_i \leq n_{i+1}$, $\bar{n} = (n_1, n_2, \ldots, n_r)$ is the type of X (n_i could be even), r is the rank of X. Let $j_1, j_2, \ldots, j_\ell$ be a sequence of integers defined by the relations $j_\ell = r$, $n_{j_i} < n_{j_{i+1}}$ and $n_{j_{i-1}} < n_k = n_{j_i}$ whenever $j_{i-1} < k \leq j_i$. $\ell = \ell(X)$ is thus the number of integers n with $QH^n(X,Q) \neq 0$. Throughout this paper let X be an H_0 space of type $\bar{n} = (n_1, \ldots, n_r)$ (and with the invariants $j_1, j_2, \ldots, j_\ell$, $\ell = \ell(X)$ derived from $\bar{n}$) and $\pi_n(X) = 0$ for $n \geq N(X)$.

1.2. <u>Theorem</u> (Toda-Mimura, see [6]): For every prime q, integer m_0 and a set of primes $\mathbb{P}_1$, $q \notin \mathbb{P}_1$, there exists $h_q : X \longrightarrow X$ so that $H^m(h_q, Z_q) = 0$ for $0 < m \leq m_0$ and $H^*(h_q, Z_p)$ is an isomorphism for $p \in \mathbb{P}_1$. Consequently, X is p-universal. It follows that for every Y, $\pi_n(Y) = 0$ for $n > N(Y)$, $Y_p \approx X_p$ if and only if there exists $f_p : Y \longrightarrow X$ with $H^*(f_p, Z_p)$ an isomorphism (such an f_p is called a p-equivalence). Consequently $[Y] \in G(X)$ if and only if for every $p \in \mathbb{P}$ ($\mathbb{P}$ the set of all primes) there exists a p-equivalence $f_p : Y \longrightarrow X$.

1.3. <u>Proposition</u>: Let $\mathbb{P}_1$ and $\mathbb{P}_2$ be disjoint sets of primes. There exists a map $m : X \times X \longrightarrow X$ so that $m|X \times *$ is a $\mathbb{P}_1$ equivalence (i.e., p-equivalence for every $p \in \mathbb{P}_1$) and $m|* \times X$ is a $\mathbb{P}_2$ equivalence.

<u>Proof</u>: Let $h : X \longrightarrow K(Z,\bar{n}) = \prod_{i=1}^{r} K(Z,n_i)$ yield an isomorphism of rational cohomology. Decompose h into a sequence $X = X_k \xrightarrow{h_k} X_{k-1} \longrightarrow \cdots X_1 \xrightarrow{h_1} X_0 = K(Z,\bar{n})$ so that $h_j : X_j \longrightarrow X_{j-1}$ is a principal $K(Z_{p_j}, s_j - 1)$ fibration induced by $k_{j-1} : X_{j-1} \longrightarrow K(Z\ \ ,s_j)$. The H-structure on X_0 yields the following (homotopy) commutative diagram for $j = 0$:

$$(1.3.1)_j \qquad \begin{array}{ccccc} X \vee X & \xrightarrow{f'_j \vee f''_j} & X \vee X & \xrightarrow{\mathscr{F}} & X \\ & & & & \downarrow h_{k,j+1} \\ \downarrow i & & & & X_{j+1} \\ & & & & \downarrow h_{j+1} \\ X \times X & & \xrightarrow{\mu_j} & & X_j \end{array}$$

$f'_0 = 1$, $f''_0 = 1$, $\mu_0 = \mu_{K(Z,\bar{n})} \circ (h \times h)$, $\mathcal{F}$ the folding map, i the inclusion, $h_{k,j+2} = h_{j+2} \circ h_{j+3} \cdots \circ h_k$. Suppose inductively $(1.3.1)_j$ exists with f'_j and f''_j being a $\mathbb{P}_1$ and a $\mathbb{P}_2$ equivalence respectively. The obstruction for lifting μ_j to $\mu_{j+1} : X \times X \longrightarrow X_{j+1}$ with $h_{k,j+2} \circ \mathcal{F} \circ (f'_j \vee f''_j) \sim \mu_{j+1} \circ i$ is a class $[\tilde{\mu}] \in [X \wedge X, K(Z_{p_{j+1}}, s_{j+1})]$ with $[\tilde{\mu}] \circ [\Lambda] = [k_j] \circ [\mu_j]$, $\Lambda : X \times X \longrightarrow X \wedge X$ the identification map. Assume without a loss of generality that $p_{j+1} \notin \mathbb{P}_1$ and let $g : X \longrightarrow X$ be a $\mathbb{P}_1$ equivalence with $H^k(g, Z_{p_{j+1}}) = 0$ for $k < s_{j+1}$. Replacing f'_j by $f'_{j+1} = f'_j \circ g$, μ_j by $\mu_j(g \times 1)$, $\tilde{\mu}$ by $\tilde{\mu}(g \wedge 1)$ the latter is null homotopic as $H^{s_{j+1}}(g \wedge 1, Z_{p_{j+1}}) = 0$ and $\mu_j(g \times 1)$ lifts to $\mu_{j+1} : X \times X \longrightarrow X_{j+1}$ and $(1.3.1)_{j+1}$ exists with $f'_{j+1} = f'_j \circ g$, and $f''_{j+1} = f'_j$ and 1.3 follows.

1.4. <u>Corollary</u>: Let $\{\mathbb{P}_i\}$ $i = 1,2,\dots,k$ be mutually disjoint sets of primes. There exists a map $\bar{m} : X^k = X \times X \cdots \times X \longrightarrow X$ so that $\bar{m} \circ i_j$ is a $\mathbb{P}_j$ equivalence, $i_j : X \longrightarrow X^k$ the injection of the j-th factor, $j = 1,2,\dots,k$.

1.5. <u>Proposition</u>: Let $[Y] \in G(X)$. For any finite set of primes $\mathbb{P}' = \{p_1, p_2, \dots, p_k\}$ there exists a $\mathbb{P}'$ equivalence $f' : Y \longrightarrow X$.

<u>Proof</u>: Let $\bar{m} : X^k \longrightarrow X$ be the map of 1.4 for $\mathbb{P}_i = \{p_i\}$. Let $h_i : Y \longrightarrow X$ be a $\mathbb{P}_i$ equivalence and let $g_i : Y \longrightarrow Y$ be a $\mathbb{P}_i$ equivalence with $H^n(g_i, Z_{p_j}) = 0$ for $j \neq i$, $0 < n \leq N(X)$. Then $f' = \bar{m} \circ (h_1 \times h_2 \times \cdots \times h_k) \circ (g_1 \times g_2 \times \cdots \times g_k) \circ \Delta^k$ (Δ^k the iterated diagonal) is the desired $\mathbb{P}'$ equivalence as $H^*[(\bar{m} \circ i_j \circ h_j \circ g_j), Z_{p_j}]$ is an isomorphism in $\dim \leq N(X)$.

For a finite abelian group G let $R(G)$ denote the maximal order of its elements: $G \longrightarrow G \otimes Z_k$ is an isomorphism if and only if $R(G)|k$. If $G_1 \longrightarrow G_2 \longrightarrow G_3$ is exact then $R(G_2)|R(G_1) \cdot R(G_3)$, $R(G \oplus G) = R(G)$. Let $\sigma_n : \pi_n(X) \longrightarrow P(H_n(X,Z)/\text{torsion}$ be the Hurewicz-Serre homomorphism. As X is an H_0 space $\sigma_n \otimes Q$ is an isomorphism. Let $\bar{t}_j = \bar{t}_j(X) = R(\ker \sigma_j(X)) \cdot R(\text{coker } \sigma_{j+1}(X))$. Then $\bar{t}_j$ and $\bar{t} = \bar{t}(X) = \prod_{j \leq N} \bar{t}_j$ are invariants of $G(X)$.

1.6. <u>Lemma</u>: Let $h_0 : X \longrightarrow K(Z,\bar{n}) = \prod_{j=1}^{r} K(Z, n_j)$ be any map inducing an isomorphism of $QH^*(\ ,Z)$/torsion. Then

$$R(\pi_j(\text{fiber } h_0))\,|\,\bar{t}_j(X).$$

Proof: $QH^*(f,Z)$/torsion is dual to $P(H_*(X,Z)/\text{torsion})$, hence, $P(H_*(h_0,Z)/\text{torsion})$ is an isomorphism and so is $\sigma_n(K(Z,\bar{n})) : \pi_n(K(Z,\bar{n})) \longrightarrow P(H_n(K(Z,\bar{n}),Z)/\text{torsion})$. As $\sigma_*(K(Z,\bar{n}))\ \pi_*(h_0) = P(H_*(h_0,Z)/\text{torsion})\ \sigma_*(X)$, $\ker \sigma_n(X) \approx \ker \pi_n(h_0)$, $\operatorname{coker} \sigma_n(X) \approx \operatorname{coker} \pi_n(h_0)$ and 1.6 follows from the exact sequence

$$0 \longrightarrow \operatorname{coker} \pi_{n+1}(h_0) \longrightarrow \pi_n(\text{fiber } h_0) \longrightarrow \ker \pi_n(h_0) \longrightarrow 0 .$$

1.7. Proposition: Let $x_{n_1}, x_{n_2}, \ldots, x_{n_r}$, $x_{n_i} \in H^{n_i}(X,Z)$ be any set which represents a basis for $QH^*(X,Z)$/torsion. There exists a map $\eta : X \times K(Z,\bar{n}) \longrightarrow X$ satisfying $X \xrightarrow{i_1} X \times K(Z,\bar{n}) \xrightarrow{\eta} X$ is the identity, $\mu_0(h_0 \times \lambda_{\bar{t}}) \sim h_0 \circ \eta$ where $h_0 : X \longrightarrow K(Z,\bar{n})$ realizes $\{x_{n_i}\}$, $\mu_0 : K(Z,\bar{n}) \times K(Z,\bar{n}) \longrightarrow K(Z,\bar{n})$ is the standard multiplication and $\lambda_{\bar{t}} : K(Z,\bar{n}) \longrightarrow K(Z,\bar{n})$ is the $\bar{t}$-th power map.

Proof: One studies the lifting process in

$$(1.7.1)_j \qquad \begin{array}{ccc} X & \longrightarrow & X_{j+1} \\ \downarrow{\scriptstyle i_1} & & \downarrow \\ X \times K(Z,\bar{n}) & \xrightarrow{\ \eta_j\ } & X_j \\ \downarrow & & \downarrow \\ X \times K(Z,\bar{n})/X \times * & \longrightarrow & K(\pi_{j+1}(\text{fiber } h_0),\ j+2) \end{array}$$

where $X \longrightarrow X_{j+1} \longrightarrow X_j \longrightarrow K(\pi_{j+1}(\text{fiber } h_0),\ j+2)$ are induced by the Postnikov decomposition of h_0. $X_0 = K(Z,\bar{n})$ and thus the following diagram commutes for $j = 0$

$$(1.7.2)_j \qquad \begin{array}{ccc} X \times K(Z,\bar{n}) & \xrightarrow{\ \eta_j\ } & X_j \\ \downarrow{\scriptstyle h_0 \times \lambda_{\bar{t}(j)}} & & \downarrow \\ K(Z,\bar{n}) \times K(Z,\bar{n}) & \xrightarrow{\ \mu_{K(Z,\bar{n})}\ } & K(Z,\bar{n}) \end{array}$$

$$\bar{t}(j) = \prod_{i \le j} \bar{t}_i(X) .$$

Suppose inductively $(1.7.1)_j$ and $(1.7.2)_j$ exists and are homotopy commutative as $R(\pi_{j+1}(\text{fiber } h_0))$ divides $\bar{t}_{j+1}(X)$ $\eta_j(1 \times \lambda_{\bar{t}_{j+1}(X)})$ lifts to $\eta_{j+1} : X \times K(Z,\bar{n}) \longrightarrow X_{j+1}$, $(1.7.1)_{j+1}$ and $(1.7.2)_{j+1}$ exist and commute and 1.7 follows.

1.8. Proposition: Let L, M be CW complexes, $g : L \longrightarrow M$. Given a map $f : L \longrightarrow X$, a set $\{x_{n_1}, x_{n_2}, \ldots, x_{n_r}\}$, $x_{n_i} \in H^{n_i}(X,Z)$ which represent a basis for $QH^*(X,Z)$/torsion and an arbitrary set $\{z_{n_1}, z_{n_2}, \ldots, z_{n_r}\}$, $z_{n_i} \in H^{n_i}(M,Z)$. There exists $f' : L \longrightarrow X$ so that for $i = 1,2,\ldots,r$

$$[H^*(f',Z) - H^*(f,Z)]\, x_{n_i} = \bar{t}(X) \cdot (H^*(g,Z)\, z_{n_i}) .$$

Moreover, $\pi(f')|(\text{torsion } \pi(L)) = \pi(f)|(\text{torsion } \pi(L))$ and $\pi(f') - \pi(f)$ factors through $\pi(g) : \pi(L) \longrightarrow \pi(M)$.

Proof: Let $\eta : X \times K(Z,\bar{n}) \longrightarrow X$ be the map of 1.6 with respect to the set $\{x_{n_1}, \ldots, x_{n_r}\}$. Put $f' = \eta \circ (f \times \bar{g}) \circ (1 \times g) \circ \Delta$ where $\bar{g} : M \longrightarrow K(Z,\bar{n})$ realizes $\{z_{n_i}\}$. f' is the desired map.

2. THE GROUP G(X)

Let $\bar{t} = \bar{t}(X)$ be as in 1.7 and 1.8. Let $t = t(X)$ be the smallest integer divisible by $\bar{t}$ and by all torsion primes of $H_m(X,Z)$, and $QH^m(X,Z)$, $m \le N(X)$. Put $\mathbb{P}_t = \{p \in \mathbb{P} \mid p | t\}$.

2.1. Proposition: Let $[Y] \in G(X)$ and let $f : Y \longrightarrow X$ be a $\mathbb{P}_t$ equivalence (see 1.5). There exists $f' : Y \longrightarrow X$ so that $\pi(f) = \pi(f')$ (hence f' is a $\mathbb{P}_t$ equivalence), $QH^*(f',Z)$/torsion $= QH^*(f,Z)$/torsion and so that there exists an f' related splitting

$$s(X) : QH^*(X,Z)/\text{torsion} \longrightarrow H^*(X,Z)$$

$$s(Y) : QH^*(Y,Z)/\text{torsion} \longrightarrow H^*(Y,Z),$$

$$H^*(f',Z)s(X) = s(Y)(QH^*(f',Z)/\text{torsion}).$$

Moreover, $s(X)$ could be chosen to be an arbitrary splitting of $H^*(X,Z) \longrightarrow QH^*(X,Z)$/torsion.

<u>Proof</u>: Choose bases $\tilde{x}_{n_1}, \tilde{x}_{n_2}, \dots, \tilde{x}_{n_r}$ and $\tilde{y}_{n_1}, \tilde{y}_{n_2}, \dots, \tilde{y}_{n_r}$ for $QH^*(X,Z)$/torsion and $QH^*(Y,Z)$/torsion respectively so that $QH^*(f,Z)$/torsion has a diagonal form: $(QH^*(f,Z)/\text{torsion})\,\tilde{x}_{n_i} = \lambda_i \tilde{y}_{n_i}$, $(\lambda_i, t) = 1$. Choose any representatives $x_{n_i} = s(X)\,\tilde{x}_{n_i} \in H^{n_i}(X,Z)$ and $y'_{n_i} = s'(Y)\tilde{y}_{n_i} \in H^{n_i}(Y,Z)$ of $\tilde{x}_{n_i}$ and $\tilde{y}_{n_i}$ respectively. Then $H^*(f,Z)x_{n_i} = \lambda_i y'_{n_i} + v_i$ where

$v_i \in \ker(H^*(Y,Z) \longrightarrow QH^*(Y,Z)/\text{torsion})$, i.e.: $a_i v_i = d_i$ is decomposable with a_i being a product of torsion primes of $QH^*(Y,Z) \approx QH^*(X,Z)$ divides some power of t, hence, $(a_i t, \lambda_i) = 1$. Let b_i, c_i be integers satisfying $1 + b_i a_i t = c_i \lambda_i$. Apply 1.8 for $L = Y$, $M = Y \wedge Y$, $g = \bar{\Delta}$ induced by the diagonal, $z_{n_i} = b_i \frac{t}{\bar{t}} w_i$, $H^*(\bar{\Delta}, Z)w_i = d_i$. Then f' of 1.8 satisfies: $\pi(f) - \pi(f')$ factors through $\pi(\bar{\Delta}) = 0$, hence $\pi(f) = \pi(f')$ and $(H^*(f',Z) - H^*(f,Z))x_{n_i} = b_i t d_i = b_i a_i t v_i$.

$H^*(f',Z)x_{n_i} = \lambda_i y'_{n_i} + v_i + b_i a_i t v_i = \lambda_i (y'_{n_i} + c_i v_i)$. $s(X)$ and $s(Y)$ defined by $s(Y)\,\tilde{y}_{n_i} = y'_{n_i} + c_i v_i = y_{n_i}$ are the desired splittings.

2.2. <u>Corollary</u>: Let $[Y] \in G(X)$ and let $h_0 : X \longrightarrow K(Z, \bar{n})$ be any map yielding an isomorphism of $QH^*(\ ,Z)$/torsion. There exists a commutative diagram

(2.2.1)
$$\begin{array}{ccc} Y & \xrightarrow{\ f_1\ } & X \\ \Big\downarrow{\scriptstyle h_1} & & \Big\downarrow{\scriptstyle h_0} \\ K(Z,\bar{n}) & \xrightarrow{\ f_0\ } & K(Z,\bar{n}) \end{array}$$

where h_0, h_1 are $\mathbb{P} - \mathbb{P}_t$ equivalences and f_0, f_1 are $\mathbb{P}_t$ equivalences, hence, Y is the pull back of h_0 and f_0. Moreover, one may assume that f_0 has a matrix form: $H^*(f_0, Z)\iota_{n_i} = \sum_{j=1}^{r} A_{ij}\iota_{n_j}$, $\det(A_{ij})$ is prime to t.

<u>Proof</u>: Let $f_1 = f'$ of 2.1. $QH^*(X,Z)$/torsion is isomorphic to the free abelian group on the fundamental classes $\iota_{n_j} \in H^{n_j}(K(Z,n_j),Z)$ and so is $QH^*(Y,Z)$/torsion. If h_0 and h_1 represent $s(X)$ and $s(Y)$ of 2.1 respectively (and as $s(X)$ can be chosen arbitrarily the only restriction on h_0 is that $QH^*(h_0,Z)$/torsion is an isomorphism) and if f_0 is chosen in a matrix form to represent $QH^*(f_1,Z)$/torsion

the commutativity of the diagram 2.2.1 is a direct consequence of 2.1. The fact that Y is the pull back follows from Zabrodsky [11], lemma 1.6. The converse of 2.2 is given by

2.3. <u>Proposition:</u> Let $h_0 : X \longrightarrow K(Z,\bar{n})$ yield an isomorphism of $QH^*(\ ,Z)$/torsion. Given any $\mathbb{P}_t$ equivalence $f_0 : K(Z,\bar{n}) \longrightarrow K(Z,\bar{n})$ form the pull back

$$\begin{array}{ccc} Y & \xrightarrow{f_1} & X \\ \Big\downarrow{\scriptstyle h_1} & & \Big\downarrow{\scriptstyle h_0} \\ K(Z,\bar{n}) & \xrightarrow{f_0} & K(Z,\bar{n}) \end{array} .$$

Then $[Y] \in G(X)$ and $QY^*(h_1,Z)$/torsion is an isomorphism.

<u>Proof:</u> As f_0 is a $\mathbb{P}_t$ equivalence so is f_1, hence $Y \approx_{\mathbb{P}_t} X$. h_i are $\mathbb{P} - \mathbb{P}_t$ equivalences, hence, $Y \approx_{\mathbb{P}-\mathbb{P}_t} K(Z,\bar{n}) \approx_{\mathbb{P}-\mathbb{P}_t} X$ and $[Y] \in G(X)$. As h_1 is a $\mathbb{P} - \mathbb{P}_t$ equivalence $QH^*(h_1,Z)$/torsion is a $\mathbb{P} - \mathbb{P}_t$ isomorphism. As $QH^*(h_0,Z)$/torsion is an isomorphism and $QH^*(f_i,Z)$/torsion $i = 0,1$ are $\mathbb{P}_t$ isomorphism $QH^*(h_1,Z)$/torsion is a $\mathbb{P}_t$ isomorphism and hence an isomorphism.

Now, fix $h_0 : X \longrightarrow K(Z,\bar{n})$ so that $QH^*(h_0,Z)$/torsion is an isomorphism. Let $\mathcal{M}_t(Z,\bar{n})$ be the set of all matrices representing End[$QH^*(K(Z,\bar{n}),Z)$/torsion] whose determinants are prime to t. For every $A \in \mathcal{M}_t(Z,\bar{n})$ let $f_0 = f_0(A) : K(Z,\bar{n}) \longrightarrow K(Z,\bar{n})$ be given by $H^*(f_0,Z)\iota_{n_i} = \sum_{j=1}^{r} A_{ij}\iota_{n_j}$, $A = (A_{ij})$.

By 2.3 the pull back of h_0 and $f_0(A)$ yields an element $\xi'(A) \in G(X)$. Thus, by 2.2 and 2.3 one has

2.4. <u>Proposition:</u> There exists a function $\xi' : \mathcal{M}_t(Z,\bar{n}) \longrightarrow G(X)$ which is onto

Now, if $[Y] \in G(X)$, $[Y] = \xi'(A)$ and if $f_1 : Y \longrightarrow X$ is a $\mathbb{P}_t$ equivalence that covers $f_0(A) : K(Z,\bar{n}) \longrightarrow K(Z,\bar{n})$, by 1.8 for any $B \in$ End[$QH^*(X,Z)$/torsion] there exists $f_B : Y \longrightarrow X$ with $QH^*(f_B,Z)$/torsion $= A + tB$ and torsion $\pi(f_A) =$ $=$ torsion $\pi(f_B)$. Checking the commutative diagram relating $\pi(f_B)$/torsion, $PH_*(f_B,Z)$/torsion, $\tilde{\sigma}(Y)$ and $\tilde{\sigma}(X)$, $\tilde{\sigma} : \pi(\ )$/torsion $\longrightarrow P(H_*(\ ,Z)$/torsion) (note that $\det \tilde{\sigma}(Y) = \det \tilde{\sigma}(X) \neq 0$ it follows that $\det(\pi(f_B)/\text{torsion}) = \det(A + tB)$ which is prime to t. Hence $\pi(f_B)/\text{torsion}$ as well as torsion $\pi(f_B) = \pi(f_A)$ are $\mathbb{P}_t$ isomorphisms, hence so is $\pi(f_B)$ and f_B is a $\mathbb{P}_t$ equivalence. By 2.2 and 2.3 $[Y] = \xi'(A + tB)$. Thus:

2.5. Proposition: If $A, A' \in \mathcal{M}_t(Z,\bar{n})$, $A - A' = tB$ then $\xi(A) = \xi(A')$. Hence ξ' yields a function $\xi'' : \mathcal{M}_t(Z,\bar{n}) \otimes Z_t = GL(Z_t,\bar{n}) \longrightarrow G(X)$ which is onto. $GL(Z_t,\bar{n}) = GL(Z_t,n_{j_1}) \oplus GL(Z_t,n_{j_2} - n_{j_1}) \oplus \cdots \oplus GL(Z_t,n_{j_\ell} - n_{j_{\ell-1}})$. Finally, if $E \in GL(\bar{n},Z)$ it can be easily seen that $\xi'(EA) = \xi'(A)$. Now, one has an exact sequence

$$GL(\bar{n},Z) \longrightarrow GL(\bar{n},Z_t) \xrightarrow{|\det|} (Z_t^*/\pm 1)^\ell \longrightarrow 0$$

so that one obtains

2.6. Proposition: There exists a function $\xi : (Z_t^*/\pm 1)^\ell \longrightarrow G(X)$ which is onto. ξ is given as follows: If $d_1, d_2, \ldots, d_\ell \in Z$, $(d_i,t) = 1$ let $A = A(d_1,d_2,\ldots,d_\ell) \in \mathcal{M}_t(Z,\bar{n})$ be the diagonal matrix with $A_{n_{j_i},n_{j_i}} = d_i$, $A_{n_k,n_k} = 1$ if $n_{j_{i-1}} < k < n_{j_i}$. Then $\xi(\bar{d}_1,\bar{d}_2,\ldots,\bar{d}_\ell) = \xi'(A)$, ξ' as in 2.4, $\bar{d}_i$ is the class of d_i in $Z_t^*/\pm 1$.

To complete the proof of the main theorem one has to show that $\ker \xi = \operatorname{im} \alpha$, $\alpha : [X,X]_t \longrightarrow (Z_t^*/\pm 1)^\ell$, i.e.: $\xi(\bar{d}) = \xi(\bar{d}')$ implies $\bar{d}' \cdot \bar{d}^{-1} \in \operatorname{im} \alpha$.

Suppose $[Y] = \xi(\bar{d}) = \xi(\bar{d}')$. Let

$$\begin{array}{ccc} Y & \xrightarrow{f_1} & X \\ \downarrow{\scriptstyle h_1} & & \downarrow{\scriptstyle h_0} \\ K(Z,\bar{n}) & \xrightarrow{f_0(\bar{A})} & K(Z,\bar{n}) \end{array} \qquad\qquad \begin{array}{ccc} Y & \xrightarrow{f_1'} & X \\ \downarrow{\scriptstyle h_1} & & \downarrow{\scriptstyle h_0} \\ K(Z,\bar{n}) & \xrightarrow{f_0(\bar{A}')} & K(Z,\bar{n}) \end{array}$$

be the representation of Y as in 2.3,

$$|\det(\bar{A} \otimes Z_t)| = \bar{d} = (d_1,d_2,\ldots,d_\ell)$$

$$|\det(\bar{A}' \otimes Z_t)| = \bar{d}' = (d_1',d_2',\ldots,d_\ell') .$$

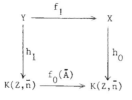

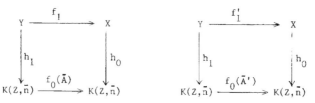

Let $\tilde{A} \in \mathcal{M}_t(Z,\bar{n})$ represent the inverse of $\bar{A}$ in $GL(\bar{n},Z_t)$, $f_0(\tilde{A}) : K(Z,\bar{n}) \longrightarrow K(Z,\bar{n})$ its geometric realization then the pull back of h_1 and $f_0(\tilde{A})$ is the pull back of h_0 and $f_0(\bar{A}) \circ f_0(\tilde{A}) = f_0(\tilde{A}\cdot\bar{A})$. As $|\det(\tilde{A}\cdot\bar{A} \otimes Z_t)| = (1,1,\dots,1)$ the pull back of h_1 and $f_0(\tilde{A})$ is in the homotopy class $\xi(1,\dots,1) = [X]$. Hence, there exists $\tilde{f}_1 : X \longrightarrow Y$ with $|\det(QH^*(f_1,Z)/\mathrm{torsion} \otimes Z_t)| = |\det(\tilde{A} \otimes Z_t)| = \bar{d}^{-1}$. $[\tilde{f}_1 \circ f_1'] \in [X,X]_t$ and $\alpha[\tilde{f}_1 \circ f_1'] = \bar{d}'\,\bar{d}^{-1}$.

The group structure of $G(X)$ can be given as follows: Let $[Y_1],[Y_2] \in G(X)$. Let $f_1 : Y_1 \longrightarrow X$ be any $\mathbb{P}_t$ equivalence. Then f_1 is a $\mathbb{P}_1$ equivalence with $\mathbb{P}_1 \supset \mathbb{P}_t$ and $\mathbb{P} - \mathbb{P}_1$ is finite. Let $f_2 : Y_2 \longrightarrow X$ be a $\mathbb{P}_t \cup (\mathbb{P} - \mathbb{P}_1)$ equivalence. Then the pull back of f_1 and f_2 represents $[Y_1] \cdot [Y_2]$.

3. NON CANCELLATION AND PRODUCTS

Let M_n be any rank 1 type (n) H_0 space with $\pi_k(X) = 0$ for $k > N(M_n)$. Suppose M_n admits a map of any degree, i.e.: $[M_n,M_n] \longrightarrow \mathrm{End}(QH^n(M_n,Z)/\mathrm{torsion})$ is surjective. E.g.: $M_n = K(Z,n)$, $M_{2n} = (\Omega S^{2n+1})_N$, $M_{2n+1} = (S^{2n+1})_N$ where $(\ )_N$ indicates the Postnikov approximation in $\dim \leq N$.

Let $X' = \prod_{i=1}^{\ell(X)} M_{n_{j_i}}$. Let $\tilde{t} = t(X \times X')$. It can be easily seen that $\alpha : [X \times X']_{\tilde{t}} \longrightarrow (Z_{\tilde{t}}^*/\pm 1)^{\ell}$ is surjective, hence $G(X \times X') = 0$. If $[Y] \in G(X)$ then $[Y \times X'] \in G(X \times X')$ and one has

3.1. <u>Non cancellation theorem:</u> If $[Y] \in G(X)$ then $Y \times X' \approx X \times X'$.

If $\tilde{X}$ is a finite dimensional H_0 space, $\tilde{X}$ of type $(n_1,n_2,\dots,n_r)$, n_i-odd, let $X = \tilde{X}_N$, $N > 2 \dim \tilde{X}$. Let $\tilde{X}' = S^{n_{j_1}} \times S^{n_{j_2}} \times \cdots \times S^{n_{j_\ell}}$, then $\dim \tilde{X}' \leq \dim \tilde{X}$. Put $X' = (\tilde{X}')_N$. By 3.1, if $[\tilde{Y}] \in G(\tilde{X})$, $\tilde{Y}_N = Y$ then $X \times X' \approx Y \times X'$. But as $\dim \tilde{X} \times \tilde{X}' = \dim \tilde{Y} \times \tilde{X}' \leq N$ one concludes that $\tilde{X} \times \tilde{X}' \approx \tilde{Y} \times \tilde{X}'$ (see Mislin [8], page 83).

3.2. <u>Remark:</u> If X is an H-space, then for every $d \in Z_t^*/\pm 1$ - $(d,d,\dots,d) \in (Z_t^*/\pm 1)^{\ell}$ is in $\mathrm{im}\,\alpha$. Hence if X'' is X' with any one of the factors of X' in 3.1 omitted $\alpha : [X \times X'', X \times X'']_{\tilde{t}} \longrightarrow (Z_t^*/\pm 1)^{\ell}$ is still surjective and $X \times X'' \approx Y \times X''$ for $[Y] \in G(X)$ and X'' has only $\ell-1$ factors. From the definition of $t = t(X)$ it follows that $t(X \times X) = t(X)$. One can easily see that if $[Y] \in G(X)$, $[Y] = \xi(d_1,\ldots,d_\ell)$ then $[Y^n] \in G(X^n)$ and $[Y^n] = \xi(d_1^n, d_2^n, \ldots, d_\ell^n)$. For $n = \varphi(t)/2 = \mathrm{order}(Z_t^*/\pm 1)$ $d_i^n = 1$ and hence, one has

3.3. <u>The product property</u>: If $[Y] \in G(X)$ then $Y^{\varphi(t)/2} \approx X^{\varphi(t)/2}$.

Converses for 3.2 and 3.3 with X an H-space were proved in Mislin [9] and Wilkerson [10].

3.4. <u>Examples</u>: E_{5w} the Hilton-Roitberg manifold of Hilton and Roitberg [4]. $[E_{5w}] \in G(Sp(2))$. As $Sp(2)$ is of type 3,7 by 3.2

$$Sp(2) \times S^7 \approx E_{5w} \times S^7 \quad \text{and} \quad Sp(2) \times S^3 \approx E_{5w} \times S^3 .$$

For $N = 10 = \dim Sp(2)$ the study of $G(Sp(2))$ and $G(Sp(2)_N)$ is completely equivalent. $t(Sp(2))$ is divisible by 2 and 3 only, $\mathbb{P}_t = \{2,3\}$. Now, (a,a) and $(1, 1+12k) \in \mathrm{im}\,\alpha \subset (Z_t^*/\pm 1)^2$, hence, one has a surjection $\bar{\xi} : (Z_{12}^*/\pm 1) \longrightarrow G(X)$. $Z_{12}^*/\pm 1 = \{1,5\}$ and as $E_{5w} \not\approx Sp(2)$ $G(X) = \{Sp(2), E_{5w}\}$. $\varphi(12)/2 = 2$ implies $E_{5w}^2 \approx Sp(2)^2$.

Let $X = G_2$. The type is 3,11. In Mimura, Nishida, and Toda [7] it was proved that $(1, 1 \pm 30k) \in \mathrm{im}\,\alpha$ as well as (a,a). Hence ξ factors through $\bar{\xi} : Z_{30}^*/\pm 1 \longrightarrow G(X)$. $(\mathbb{P}_t = \{2,3,5\})$. Now $Z_{30}^*/\pm 1 = \{1,7,11,13\}$. $i \in \ker \bar{\xi}$ implies the existence of a map of type $(1, \pm i)$ and consequently the existence of a map of type $(0, \pm i - 1)$. But one can see that if $H^3(f, Z_p) = 0$ $p = 2,3,5$ then $H^{11}(f,Z)/\mathrm{torsion} \otimes Z_p = 0$, hence $i \equiv 1 \pmod{30}$. It follows that $\ker \bar{\xi} = \{1\}$ and $G(G_2) = \{X_1 = G_2, X_7, X_{11}, X_{13}\}$ (see Hilton, Mislin and Roitberg [5]). $X_i \times S^{11} \approx G_2 \times S^{11}$ and $X_i \times S^3 \approx G_2 \times S^3$ by 3.1.

REFERENCES

[1] J. M. Cohen, Stable homotopy. Lecture Notes in Mathematics 165, Springer-Verlag, 1970.

[2] P. Freyd, Stable homotopy. Proceedings of the Conference on Categorical Algebra (La Jolla, 1965), Springer-Verlag, 1966.

[3] P. J. Hilton, On the Grothendieck group of compact polyhedra. Fund. Math. 61, 199-214 (1967).

[4] P. J. Hilton and J. Roitberg, On principal S^3 bundles over spheres. Ann. of Math. 90, 91-107 (1969).

[5] P. J. Hilton, G. Mislin and J. Roitberg, H-spaces of rank 2 and non cancellation phenomena. Inv. Math. 16, 325-334 (1972).

[6] M. Mimura and H. Toda, On p-equivalences and p-universal spaces. Comm. Math. Helv. 46, 87-97 (1971).

[7] Mimura, Nishida and Toda, H-spaces of rank 2. (Mimeographed)

[8] G. Mislin, The genus of an H-space. Symposium on Algebraic Topology (Seattle, 1971), Lecture Notes in Mathematics 249, p. 75-83. Springer-Verlag.

[9] ————, Cancellation properties of H-spaces. (To appear.)

[10] C. Wilkerson,

[11] A. Zabrodsky, On the genus of finite CW-H-spaces. Comm. Math. Helv. (To appear.)

ADDRESSES OF CONTRIBUTORS

Professor D. W. Anderson
University of California at San Diego
Department of Mathematics
La Jolla, California 92037

Professor Martin Arkowitz
Dartmouth College
Department of Mathematics
Hanover, New Hampshire 03755

Professor Martin Bendersky
University of Washington
Department of Mathematics
Seattle, Washington 98195

Dr. Richard Body
University of British Columbia
Department of Mathematics
Vancouver 8, B. C., Canada

Professor A. K. Bousfield
University of Illinois
Department of Mathematics
Chicago, Illinois 60680

Professor Morton Curtis
Rice University
Department of Mathematics
Houston, Texas 77001

Professor Aristide Deleanu
University of Syracuse
Department of Mathematics
Syracuse, New York 13210

Professor Henry Glover
Ohio State University
Department of Mathematics
Columbus, Ohio 43210

Professor John Harper
University of Rochester
Department of Mathematics
Rochester, New York 14627

Professor Peter Hilton
Battelle Seattle Research Center
4000 N. E. 41st
Seattle, Washington 98105

Professor Jeanne Meisen
Case Western Reserve University
Department of Mathematics
Cleveland, Ohio 44106

Professor Guido Mislin
Lehrstuhl für Mathematik
ETH
Zürich, Switzerland

Professor David Rector
Rice University
Department of Mathematics
Houston, Texas 77001

Professor Joseph Roitberg
Institute for Advanced Study
Princeton, New Jersey 08540

Professor James Stasheff
Temple University
Department of Mathematics
Philadelphia, Pennsylvania 19122

Professor Clarence Wilkerson
Carleton University
Department of Mathematics
Ottawa, Canada K1S 5B6

Professor Alexander Zabrodsky
Institute for Advanced Study
Princeton, New Jersey 08540

Vol. 247: Lectures on Operator Algebras. Tulane University Ring and Operator Theory Year, 1970–1971. Volume II. XI, 786 pages. 1972. DM 40,–

Vol. 248: Lectures on the Applications of Sheaves to Ring Theory. Tulane University Ring and Operator Theory Year, 1970–1971. Volume III. VIII, 315 pages. 1971. DM 26,–

Vol. 249: Symposium on Algebraic Topology. Edited by P. J. Hilton. VII, 111 pages. 1971. DM 16,–

Vol. 250: B. Jónsson, Topics in Universal Algebra. VI, 220 pages. 1972. DM 20,–

Vol. 251: The Theory of Arithmetic Functions. Edited by A. A. Gioia and D. L. Goldsmith VI, 287 pages. 1972. DM 24,–

Vol. 252: D. A. Stone, Stratified Polyhedra. IX, 193 pages. 1972. DM 18,–

Vol. 253: V. Komkov, Optimal Control Theory for the Damping of Vibrations of Simple Elastic Systems. V, 240 pages. 1972. DM 20,–

Vol. 254: C. U. Jensen, Les Foncteurs Dérivés de $\varprojlim$ et leurs Applications en Théorie des Modules. V, 103 pages. 1972. DM 16,–

Vol. 255: Conference in Mathematical Logic – London '70. Edited by W. Hodges. VIII, 351 pages. 1972. DM 26,–

Vol. 256: C. A. Berenstein and M. A. Dostal, Analytically Uniform Spaces and their Applications to Convolution Equations. VII, 130 pages. 1972. DM 16,–

Vol. 257: R. B. Holmes, A Course on Optimization and Best Approximation. VIII, 233 pages. 1972. DM 20,–

Vol. 258: Séminaire de Probabilités VI. Edited by P. A. Meyer. VI, 253 pages. 1972. DM 22,–

Vol. 259: N. Moulis, Structures de Fredholm sur les Variétés Hilbertiennes. V, 123 pages. 1972. DM 16,–

Vol. 260: R. Godement and H. Jacquet, Zeta Functions of Simple Algebras. IX, 188 pages. 1972. DM 18,–

Vol. 261: A. Guichardet, Symmetric Hilbert Spaces and Related Topics. V, 197 pages. 1972. DM 18,–

Vol. 262: H. G. Zimmer, Computational Problems, Methods, and Results in Algebraic Number Theory. V, 103 pages. 1972. DM 16,–

Vol. 263: T. Parthasarathy, Selection Theorems and their Applications. VII, 101 pages. 1972. DM 16,–

Vol. 264: W. Messing, The Crystals Associated to Barsotti-Tate Groups: With Applications to Abelian Schemes. III, 190 pages. 1972. DM 18,–

Vol. 265: N. Saavedra Rivano, Catégories Tannakiennes. II, 418 pages. 1972. DM 26,–

Vol. 266: Conference on Harmonic Analysis. Edited by D. Gulick and R. L. Lipsman. VI, 323 pages. 1972. DM 24,–

Vol. 267: Numerische Lösung nichtlinearer partieller Differential- und Integro-Differentialgleichungen. Herausgegeben von R. Ansorge und W. Törnig, VI, 339 Seiten. 1972. DM 26,–

Vol. 268: C. G. Simader, On Dirichlet's Boundary Value Problem. IV, 238 pages. 1972. DM 20,–

Vol. 269: Théorie des Topos et Cohomologie Etale des Schémas. (SGA 4). Dirigé par M. Artin, A. Grothendieck et J. L. Verdier. XIX, 525 pages. 1972. DM 50,–

Vol. 270: Théorie des Topos et Cohomologie Etale des Schémas. Tome 2. (SGA 4). Dirigé par M. Artin, A. Grothendieck et J. L. Verdier. V, 418 pages. 1972. DM 50,–

Vol. 271: J. P. May, The Geometry of Iterated Loop Spaces. IX, 175 pages. 1972. DM 18,–

Vol. 272: K. R. Parthasarathy and K. Schmidt, Positive Definite Kernels, Continuous Tensor Products, and Central Limit Theorems of Probability Theory. VI, 107 pages. 1972. DM 16,–

Vol. 273: U. Seip, Kompakt erzeugte Vektorräume und Analysis. IX, 119 Seiten. 1972. DM 16,–

Vol. 274: Toposes, Algebraic Geometry and Logic. Edited by. F. W. Lawvere. VI, 189 pages. 1972. DM 18,–

Vol. 275: Séminaire Pierre Lelong (Analyse) Année 1970–1971. VI, 181 pages. 1972. DM 18,–

Vol. 276: A. Borel, Représentations de Groupes Localement Compacts. V, 98 pages. 1972. DM 16,–

Vol. 277: Séminaire Banach. Edité par C. Houzel. VII, 229 pages. 1972. DM 20,–

Vol. 278: H. Jacquet, Automorphic Forms on GL(2). Part II. XIII, 142 pages. 1972. DM 16,–

Vol. 279: R. Bott, S. Gitler and I. M. James, Lectures on Algebraic and Differential Topology. V, 174 pages. 1972. DM 18,–

Vol. 280: Conference on the Theory of Ordinary and Partial Differential Equations. Edited by W. N. Everitt and B. D. Sleeman. XV, 367 pages. 1972. DM 26,–

Vol. 281: Coherence in Categories. Edited by S. Mac Lane. VII, 235 pages. 1972. DM 20,–

Vol. 282: W. Klingenberg und P. Flaschel, Riemannsche Hilbertmannigfaltigkeiten. Periodische Geodätische. VII, 211 Seiten. 1972. DM 20,–

Vol. 283: L. Illusie, Complexe Cotangent et Déformations II. VII, 304 pages. 1972. DM 24,–

Vol. 284: P. A. Meyer, Martingales and Stochastic Integrals I. VI, 89 pages. 1972. DM 16,–

Vol. 285: P. de la Harpe, Classical Banach-Lie Algebras and Banach-Lie Groups of Operators in Hilbert Space. III, 160 pages. 1972. DM 16,–

Vol. 286: S. Murakami, On Automorphisms of Siegel Domains. V, 95 pages. 1972. DM 16,–

Vol. 287: Hyperfunctions and Pseudo-Differential Equations. Edited by H. Komatsu. VII, 529 pages. 1973. DM 36,–

Vol. 288: Groupes de Monodromie en Géométrie Algébrique. (SGA 7 I). Dirigé par A. Grothendieck. IX, 523 pages. 1972. DM 50,–

Vol. 289: B. Fuglede, Finely Harmonic Functions. III, 188. 1972. DM 18,–

Vol. 290: D. B. Zagier, Equivariant Pontrjagin Classes and Applications to Orbit Spaces. IX, 130 pages. 1972. DM 16,–

Vol. 291: P. Orlik, Seifert Manifolds. VIII, 155 pages. 1972. DM 16,–

Vol. 292: W. D. Wallis, A. P. Street and J. S. Wallis, Combinatorics: Room Squares, Sum-Free Sets, Hadamard Matrices. V, 508 pages. 1972. DM 50,–

Vol. 293: R. A. DeVore, The Approximation of Continuous Functions by Positive Linear Operators. VIII, 289 pages. 1972. DM 24,–

Vol. 294: Stability of Stochastic Dynamical Systems. Edited by R. F. Curtain. IX, 332 pages. 1972. DM 26,–

Vol. 295: C. Dellacherie, Ensembles Analytiques, Capacités, Mesures de Hausdorff. XII, 123 pages. 1972. DM 16,–

Vol. 296: Probability and Information Theory II. Edited by M. Behara, K. Krickeberg and J. Wolfowitz. V, 223 pages. 1973. DM 20,–

Vol. 297: J. Garnett, Analytic Capacity and Measure. IV, 138 pages. 1972. DM 16,–

Vol. 298: Proceedings of the Second Conference on Compact Transformation Groups. Part 1. XIII, 453 pages. 1972. DM 32,–

Vol. 299: Proceedings of the Second Conference on Compact Transformation Groups. Part 2. XIV, 327 pages. 1972. DM 26,–

Vol. 300: P. Eymard, Moyennes Invariantes et Représentations Unitaires. II, 113 pages. 1972. DM 16,–

Vol. 301: F. Pittnauer, Vorlesungen über asymptotische Reihen. VI, 186 Seiten. 1972. DM 18,–

Vol. 302: M. Demazure, Lectures on p-Divisible Groups. V, 98 pages. 1972. DM 16,–

Vol. 303: Graph Theory and Applications. Edited by Y. Alavi, D. R. Lick and A. T. White. IX, 329 pages. 1972. DM 26,–

Vol. 304: A. K. Bousfield and D. M. Kan, Homotopy Limits, Completions and Localizations. V, 348 pages. 1972. DM 26,–

Vol. 305: Théorie des Topos et Cohomologie Etale des Schémas. Tome 3. (SGA 4). Dirigé par M. Artin, A. Grothendieck et J. L. Verdier. VI, 640 pages. 1973. DM 50,–

Vol. 306: H. Luckhardt, Extensional Gödel Functional Interpretation. VI, 161 pages. 1973. DM 18,–

Vol. 307: J. L. Bretagnolle, S. D. Chatterji et P.-A. Meyer, Ecole d'été de Probabilités: Processus Stochastiques. VI, 198 pages. 1973. DM 20,–

Vol. 308: D. Knutson, λ-Rings and the Representation Theory of the Symmetric Group. IV, 203 pages. 1973. DM 20,–

Vol. 309: D. H. Sattinger, Topics in Stability and Bifurcation Theory. VI, 190 pages. 1973. DM 18,–

Vol. 310: B. Iversen, Generic Local Structure of the Morphisms in Commutative Algebra. IV, 108 pages. 1973. DM 16,-

Vol. 311: Conference on Commutative Algebra. Edited by J. W. Brewer and E. A. Rutter. VII, 251 pages. 1973. DM 22,-

Vol. 312: Symposium on Ordinary Differential Equations. Edited by W. A. Harris, Jr. and Y. Sibuya. VIII, 204 pages. 1973. DM 22,-

Vol. 313: K. Jörgens and J. Weidmann, Spectral Properties of Hamiltonian Operators. III, 140 pages. 1973. DM 16,-

Vol. 314: M. Deuring, Lectures on the Theory of Algebraic Functions of One Variable. VI, 151 pages. 1973. DM 16,-

Vol. 315: K. Bichteler, Integration Theory (with Special Attention to Vector Measures). VI, 357 pages. 1973. DM 26,-

Vol. 316: Symposium on Non-Well-Posed Problems and Logarithmic Convexity. Edited by R. J. Knops. V, 176 pages. 1973. DM 18,-

Vol. 317: Séminaire Bourbaki – vol. 1971/72. Exposés 400–417. IV, 361 pages. 1973. DM 26,-

Vol. 318: Recent Advances in Topological Dynamics. Edited by A. Beck, VIII, 285 pages. 1973. DM 24,-

Vol. 319: Conference on Group Theory. Edited by R. W. Gatterdam and K. W. Weston. V, 188 pages. 1973. DM 18,-

Vol. 320: Modular Functions of One Variable I. Edited by W. Kuyk. V, 195 pages. 1973. DM 18,-

Vol. 321: Séminaire de Probabilités VII. Edité par P. A. Meyer. VI, 322 pages. 1973. DM 26,-

Vol. 322: Nonlinear Problems in the Physical Sciences and Biology. Edited by I. Stakgold, D. D. Joseph and D. H. Sattinger. VIII, 357 pages. 1973. DM 26,-

Vol. 323: J. L. Lions, Perturbations Singulières dans les Problèmes aux Limites et en Contrôle Optimal. XII, 645 pages. 1973. DM 42,-

Vol. 324: K. Kreith, Oscillation Theory. VI, 109 pages. 1973. DM 16,-

Vol. 325: Ch.-Ch. Chou, La Transformation de Fourier Complexe et L'Equation de Convolution. IX, 137 pages. 1973. DM 16,-

Vol. 326: A. Robert, Elliptic Curves. VIII, 264 pages. 1973. DM 22,-

Vol. 327: E. Matlis, 1-Dimensional Cohen-Macaulay Rings. XII, 157 pages. 1973. DM 18,-

Vol. 328: J. R. Büchi and D. Siefkes, The Monadic Second Order Theory of All Countable Ordinals. VI, 217 pages. 1973. DM 20,-

Vol. 329: W. Trebels, Multipliers for (C, α)-Bounded Fourier Expansions in Banach Spaces and Approximation Theory. VII, 103 pages. 1973. DM 16,-

Vol. 330: Proceedings of the Second Japan-USSR Symposium on Probability Theory. Edited by G. Maruyama and Yu. V. Prokhorov. VI, 550 pages. 1973. DM 36,-

Vol. 331: Summer School on Topological Vector Spaces. Edited by L. Waelbroeck. VI, 226 pages. 1973. DM 20,-

Vol. 332: Séminaire Pierre Lelong (Analyse) Année 1971-1972. V, 131 pages. 1973. DM 16,-

Vol. 333: Numerische, insbesondere approximationstheoretische Behandlung von Funktionalgleichungen. Herausgegeben von R. Ansorge und W. Törnig. VI, 296 Seiten. 1973. DM 24,-

Vol. 334: F. Schweiger, The Metrical Theory of Jacobi-Perron Algorithm. V, 111 pages. 1973. DM 16,-

Vol. 335: H. Huck, R. Roitzsch, U. Simon, W. Vortisch, R. Walden, B. Wegner und W. Wendland, Beweismethoden der Differentialgeometrie im Großen. IX, 159 Seiten. 1973. DM 18,-

Vol. 336: L'Analyse Harmonique dans le Domaine Complexe. Edité par E. J. Akutowicz. VIII, 169 pages. 1973. DM 18,-

Vol. 337: Cambridge Summer School in Mathematical Logic. Edited by A. R. D. Mathias and H. Rogers. IX, 660 pages. 1973. DM 42,-

Vol. 338: J. Lindenstrauss and L. Tzafriri, Classical Banach Spaces. IX, 243 pages. 1973. DM 22,-

Vol. 339: G. Kempf, F. Knudsen, D. Mumford and B. Saint-Donat, Toroidal Embeddings I. VIII, 209 pages. 1973. DM 20,-

Vol. 340: Groupes de Monodromie en Géométrie Algébrique. (SGA 7 II). Par P. Deligne et N. Katz. X, 438 pages. 1973. DM 40,-

Vol. 341: Algebraic K-Theory I, Higher K-Theories. Edited by H. Bass. XV, 335 pages. 1973. DM 26,-

Vol. 342: Algebraic K-Theory II, "Classical" Algebraic K-Theory, and Connections with Arithmetic. Edited by H. Bass. XV, 527 pages. 1973. DM 36,-

Vol. 343: Algebraic K-Theory III, Hermitian K-Theory and Geometric Applications. Edited by H. Bass. XV, 572 pages. 1973. DM 38,-

Vol. 344: A. S. Troelstra (Editor), Metamathematical Investigation of Intuitionistic Arithmetic and Analysis. XVII, 485 pages. 1973. DM 34,-

Vol. 345: Proceedings of a Conference on Operator Theory. Edited by P. A. Fillmore. VI, 228 pages. 1973. DM 20,-

Vol. 346: Fučík et al., Spectral Analysis of Nonlinear Operators. II, 287 pages. 1973. DM 26,-

Vol. 347: J. M. Boardman and R. M. Vogt, Homotopy Invariant Algebraic Structures on Topological Spaces. X, 257 pages. 1973. DM 22,-

Vol. 348: A. M. Mathai and R. K. Saxena, Generalized Hypergeometric Functions with Applications in Statistics and Physical Sciences. VII, 314 pages. 1973. DM 26,-

Vol. 349: Modular Functions of One Variable II. Edited by W. Kuyk and P. Deligne. V, 598 pages. 1973. DM 38,-

Vol. 350: Modular Functions of One Variable III. Edited by W. Kuyk and J.-P. Serre. V, 350 pages. 1973. DM 26,-

Vol. 351: H. Tachikawa, Quasi-Frobenius Rings and Generalizations. XI, 172 pages. 1973. DM 18,-

Vol. 352: J. D. Fay, Theta Functions on Riemann Surfaces. V, 137 pages. 1973. DM 16,-

Vol. 353: Proceedings of the Conference. on Orders, Group Rings and Related Topics. Organized by J. S. Hsia, M. L. Madan and T. G. Ralley. X, 224 pages. 1973. DM 20,-

Vol. 354: K. J. Devlin, Aspects of Constructibility. XII, 240 pages. 1973. DM 22,-

Vol. 355: M. Sion, A Theory of Semigroup Valued Measures. V, 140 pages. 1973. DM 16,-

Vol. 356: W. L. J. van der Kallen, Infinitesimally Central-Extensions of Chevalley Groups. VII, 147 pages. 1973. DM 16,-

Vol. 357: W. Borho, P. Gabriel und R. Rentschler, Primideale in Einhüllenden auflösbarer Lie-Algebren. V, 182 Seiten. 1973. DM 18,-

Vol. 358: F. L. Williams, Tensor Products of Principal Series Representations. VI, 132 pages. 1973. DM 16,-

Vol. 359: U. Stammbach, Homology in Group Theory. VIII, 183 pages. 1973. DM 18,-

Vol. 360: W. J. Padgett and R. L. Taylor, Laws of Large Numbers for Normed Linear Spaces and Certain Fréchet Spaces. VI, 111 pages. 1973. DM 16,-

Vol. 361: J. W. Schutz, Foundations of Special Relativity: Kinematic Axioms for Minkowski Space Time. XX, 314 pages. 1973. DM 26,-

Vol. 362: Proceedings of the Conference on Numerical Solution of Ordinary Differential Equations. Edited by D. Bettis. VIII, 490 pages. 1974. DM 34,-

Vol. 363: Conference on the Numerical Solution of Differential Equations. Edited by G. A. Watson. IX, 221 pages. 1974. DM 20,-

Vol. 364: Proceedings on Infinite Dimensional Holomorphy. Edited by T. L. Hayden and T. J. Suffridge. VII, 212 pages. 1974. DM 20,-

Vol. 365: R. P. Gilbert, Constructive Methods for Elliptic Equations. VII, 397 pages. 1974. DM 26,-

Vol. 366: R. Steinberg, Conjugacy Classes in Algebraic Groups (Notes by V. V. Deodhar). VI, 159 pages. 1974. DM 18,-

Vol. 367: K. Langmann und W. Lütkebohmert, Cousinverteilungen und Fortsetzungssätze. VI, 151 Seiten. 1974. DM 16,-

Vol. 368: R. J. Milgram, Unstable Homotopy from the Stable Point of View. V, 109 pages. 1974. DM 16,-

Vol. 369: Victoria Symposium on Nonstandard Analysis. Edited by A. Hurd and P. Loeb. XVIII, 339 pages. 1974. DM 26,-

Vol. 370: B. Mazur and W. Messing, Universal Extensions and One Dimensional Crystalline Cohomology. VII, 134 pages. 1974. DM 16,-